THEY TRIED TO WARN US

OTHER BOOKS BY RAY WELLING, PhD

Digital Disruption and Transformation: Lessons from History (with Simon van Wyk)

Teach Advertising and Marketing With a Sense of Humor: Why (and How to) Be a Funnier and More Effective Advertising & Marketing Educator (with various authors)

The Men's Room: A thinking man's guide for women of the next millennium (with Toby Green)

Byline for the Dead: A Novel

THEY TRIED TO WARN US

Voices from the past ponder the world today

By Ray Welling

Published by Clear as Mud Productions, Sydney Australia

© 2025 Ray Welling

All rights reserved. No part of this publication may be reproduced, stored in a retrieval system, or transmitted in any form or by any means, electronic, mechanical, photocopying, recording or otherwise, without the prior written permission from the publisher.

Disclaimer: Every effort has been made to ensure that this book is free from error or omissions. Information provided is of general nature only and should not be considered legal or financial advice. The intent is to offer a variety of information to the reader. However, the author, publisher, editor or their agents or representatives shall not accept responsibility for any loss or inconvenience caused to a person or organization relying on this information.

A catalogue record for this book is available from the National Library of Australia.

Book cover design by Janelle Bray

ISBN:

978-0-6482711-2-3 (pbk)

978-0-6482711-3-0 (ebk)

Table Of Contents

Introduction _____ 7

Arc 1: The Medium Is the Message _____ 11

1: Neil Postman - The Prophet of the Screen _____ 13

2. Marshall McLuhan – The Medium Prophet _____ 33

3: Guy Debord – The Society of the Spectacle _____ 43

4: David Foster Wallace: The Cost of Consciousness _____ 51

5: Philip K. Dick: The Simulation Has Crashed _____ 59

6: George Orwell: The Ministry Is Real _____ 67

Arc 2: Machines, Markets, and the Loss of Meaning _____ 75

7: Herbert Marcuse: The Machine That Sells Obedience _____ 77

8: Jacques Ellul – Prophet of Technological Limits _____ 85

9: Norbert Wiener: Feedback Loop from Hell _____ 93

10: Erich Fromm – To Have or To Be _____ 103

11: Ivan Illich – Tools For Conviviality _____ 109

12: Albert Borgmann – The Burden of Ease _____ 115

Arc 3: Power, Politics, and the Algorithmic State _____ 121

13: Hannah Arendt – The Algorithmic Banality of Evil _____ 123

14: Simone Weil – Gravity, Grace, and the Machine _____ 131

15: Edward Bellamy – Looking Backward, and Seeing a Mess __ 137

16: Lewis Mumford – The Megamachine _____ 145

17: Rachel Carson – The Profits of Denial _____ 151

18: Buckminster Fuller – Final Boarding Call for Spaceship Earth __ 159

Arc 4: Myth, Meaning, and the Human Soul _____ 165

19: Mary Shelley – Frankenstein and the Ethics of Creation _____ 167

20: Jim Morrison – The End of the Endless Scroll _____ 175

21: Ursula K. Le Guin – Dreams Beyond the Machine _____ 181

22: William Blake – Silicon Babylon _____ 187

23: Alan Turing – The Machines That Imitate Us _____ 193

24: Jane Jacobs – The Soul of the Sidewalk _____ 199

25: Aldous Huxley – The Age of Engineered Pleasure _____ 205

26: What Now? A Manifesto for the Living _____ 213

Introduction

"The future is already here – it's just not evenly distributed."
– William Gibson

They warned us. Not in slogans. Not in press releases. Not in Twitter threads or TED Talks.

They warned us in books, essays, interviews, and thought experiments. They warned us with metaphors and manifestos, parables and polemics. And for the most part, we ignored them.

This book is about those voices. They came from across centuries and continents: poets, scientists, philosophers, engineers, journalists, prophets. Some were household names. Others were dismissed, marginalized, or forgotten.

What united them wasn't ideology or discipline, but vision. Each of them looked beyond the surface of their time – and saw the future rushing toward us. And then they tried to stop it.

The Premise

They Tried to Warn Us began as a podcast (which you can find at: https://open.spotify.com/show/3IlsQV7KXuaseN1d3n7Hix). A thought experiment. What if we could bring back the thinkers who saw what was coming – and ask them what they make of our world today?

Starting out as a project investigating the work of media theorist Neil Postman, it grew into a series of fictional interviews with (dead) thinkers about technology, each rooted in real philosophy and historical context. Each episode featured the host – Ray – welcoming a long-dead guest into the chaos of the 21st century. They step into our headlines.

They browse our apps. They ride our elevators. They sit across from us, squinting at the digital noise – and speak.

Sometimes with awe. Sometimes with anger. Always with urgency.

Each chapter in this book is one of those conversations (plus several other interviews in the same style beyond the podcast episodes), followed by commentary and curated references. Our goal is not to glorify the past, but to recover a clarity we've lost.

Because the truth is: we've been told.

The Pattern We Refuse to See

The thinkers in this collection aren't united by era, politics, or nationality. Some were wildly optimistic, others deeply cynical. But they each saw a dimension of modern life that most people overlooked – and they raised the alarm.

1. **Neil Postman** warned us that entertainment would swallow public discourse.

2. **Marshall McLuhan** taught us that every new medium reshapes how we think.

3. **Ivan Illich** showed how systems built to help us often end up hurting us.

4. **Rachel Carson** sounded the ecological alarm before climate change had a name.

5. **Jane Jacobs** stood between a bulldozer and a neighborhood – and won.

6. **George Orwell** feared a boot stomping the human face forever.

7. **Aldous Huxley** feared we would laugh our way into slavery.

8. **Philip K. Dick** warned that reality would become negotiable – and monetized.

9. **Mary Shelley** foresaw the ethical nightmare of creation without conscience.

These voices weren't perfect. But they weren't passive, either. They asked the questions we're still afraid to ask:

- What kind of world are we building?
- Who is being left behind?
- Are we shaping our tools – or are they shaping us?
- Are we replacing meaning with convenience?
- And at what point does progress become peril?

Why Now?

We live in an age of acceleration.

Faster computing. Bigger data. Smarter machines. Shorter attention spans. Every month brings a new app, a new feature, a new convenience. But we remain unmoored.

AI can write your résumé, compose your lullaby, and generate your wedding vows. But we still haven't figured out how to live together, how to govern wisely, or how to protect the planet beneath our feet.

We scroll. We stream. We automate. We cope. But beneath it all, the questions remain. The deeper ones. The ones these thinkers refused to let die.

This book doesn't offer easy answers. It's not a manifesto, nor a survival guide. It's a chorus of warnings – some whispered, some shouted – about the paths we've taken and the doors we've closed.

How to Read This Book

Each chapter contains the following elements:

1. A snapshot of who the thinker was, what they foresaw, and why their ideas matter in today's world.

2. A stylized dialogue between the host and guest, imagined with fidelity to their voice, tone, and philosophy.

3. A modern reflection connecting their core insights to today's cultural, political, or technological reality.

4. A list of key works (original and secondary) for deeper exploration.

You can read the book straight through, from media theorists to mystics. Or you can skip around, following your own interests. Each voice stands on its own – but taken together, they begin to harmonize.

The Hope Beneath the Warning

It would be easy to read this book as a lament. A slow funeral dirge for common sense. But if you listen closely, you'll hear something else: a quiet, persistent thread of hope.

Not optimism. Not utopianism. Hope.

Each of these thinkers believed that change was possible. That culture could be reclaimed. That technology was not destiny. That citizens – readers like you – could resist the pull of the megamachine and choose another path.

Some gave us tools. Some gave us metaphors. All of them gave us warning. And now, their voices are back. In your ears. In your hands. On this page.

The question is: Will we finally listen?

Arc 1: The Medium Is the Message

1: Neil Postman - The Prophet of the Screen

"What we need are not more gadgets but better metaphors."

Preface: The Prophet Who Was Not Amused

Neil Postman was not a futurist in the traditional sense. He didn't make sweeping predictions about flying cars or robot assistants. Instead, he observed. Carefully. Critically. And with devastating clarity.

Born in 1931 and rising to prominence as a media theorist and educator, Postman warned not about the gadgets of the future, but about how they would shape our minds, our culture, and our democracy. In his seminal 1985 book *Amusing Ourselves to Death*, he argued that television was transforming serious public discourse into entertainment. It wasn't Orwell's jackboot we needed to fear, he said, but Huxley's soma: a culture so saturated in triviality that we would forget how to think.

But Postman's vision didn't stop with television. In later works like *Technopoly* and *The End of Education*, he examined how all technologies, from computers to standardized testing, reshaped human values. He was especially concerned with the myth of technological neutrality: the idea that tools are just tools. Postman believed that every new medium had an epistemology – a way of knowing, seeing, and valuing the world. If we fail to recognize that, he warned, we will become tools of our tools.

In 2025, his warnings resonate more than ever. Social media, algorithmic news feeds, infinite scroll, ChatGPT, and the attention economy have only accelerated the transformation he described. And so we invited him back – for a fictional conversation. To ask what he sees in the world we've made.

Fictional Interview: Neil Postman in 2025

RAY:

Today's guest saw it all coming. More than 40 years ago, he warned us that the Western world was turning into a nation of screen-addicted simpletons. That our public discourse was becoming indistinguishable from daytime television. That serious thought was being replaced by snack-sized spectacle.

Neil Postman was an American educator, cultural critic, and media theorist who spent his career trying to warn us. Best known for his 1985 book *Amusing Ourselves to Death*, Postman argued that television – and later, all screen-based media – had transformed serious public discourse into a form of entertainment. In his view, the medium through which we communicate doesn't just shape the message – it shapes the way we think. In the television age, politics, news, education, even religion, all had to conform to the logic of entertainment. Substance was displaced by spectacle. Dialogue by drama. Citizenship by consumerism.

But *Amusing Ourselves to Death* was just one chapter of Postman's larger body of work. In *Technopoly: The Surrender of Culture to Technology* (1992), Neil Postman argued that technology had evolved to the point where it shaped and controlled society, rather than simply being a tool used by it.

In *The Disappearance of Childhood* (1982), he made the case that electronic media blurred the boundaries between children and adults by eliminating the distinctions once maintained through print culture. When everything is visible, instantaneous, and consumable by all ages, childhood – as a protected developmental space – begins to vanish.

And in *Teaching as a Subversive Activity* Postman challenged educators to reject rote memorization and passive learning in favor of critical inquiry. He saw schools as one of the few remaining institutions capable of resisting the conformist pressures of mass media – if only they were brave enough to try.

Postman wasn't anti-technology – but he was fiercely skeptical of any society that adopted new tools without asking, "What problem does this solve? And what does it make worse?"

So today, in the age of TikTok tutors, AI influencers, algorithmic outrage, and dopamine-driven doomscrolling, we resurrect the man who warned us about the dangers of mistaking information for wisdom – and amusement for truth.

He warned us, but we didn't listen. So we're bringing him back to get his view on what happened.

Please welcome the patron saint of cultural pessimism, media theorist Neil Postman.

POSTMAN (dry, unimpressed):

You summoned me... using a podcast? On a platform called "Spotify"? You know that name is made up, right?

RAY:

We've got worse. There's one called "X." Used to be Twitter. Now it's mostly memes and rage.

POSTMAN:

So... like the telegraph, but angry?

RAY:

Sounds pretty right. Now, for listeners born after 1985, give us the Postman Primer. Who were you, and what did you try to tell the world?

POSTMAN:

I was a media ecologist. Not a technophobe – but a technoskeptic. I studied how our tools – especially media – reshape how we think. My central argument in *Amusing Ourselves to Death* was that television had turned everything – politics, religion, journalism – into show business.

When a culture's dominant mode of communication favors entertainment over reason, serious public conversation dies.

RAY:

So not George Orwell's dystopia from 1984, where truth is forcibly suppressed by authoritarian control, but Aldous Huxley's version from Brave New World, where truth is drowned in a sea of trivia, distraction, and pleasure. From memory, you argued that we wouldn't need Big Brother to silence us, because we'd be too busy watching game shows and cat videos to notice we'd been silenced.

POSTMAN:

Exactly. Orwell feared a world of censorship. I feared a world where no one *wanted* to read. Where information was so fast, so shallow, and so plentiful, that no one would care whether it was true.

RAY:

Spoiler alert: you were right.

POSTMAN:

In Amusing Ourselves to Death, I argued that television doesn't just deliver content – it changes the form of discourse itself. It turns everything into entertainment. Politics becomes theater. Religion becomes spectacle. News becomes a series of disconnected images, stripped of context and consequence.

RAY:

And now we've got TikTok sermons and presidential debates with emojis.

POSTMAN:

Exactly. Television began the process. But your digital platforms perfected it. You've moved from typography to videography – from logic to aesthetics. And in doing so, you've traded depth for dopamine.

Television didn't just change the content of our conversations; it changed their very nature. When information is packaged for entertainment, politics, religion, education – even the news – become performances. The "Now…this" culture I described, where every story is disconnected from the last, is now the default mode of communication. We live in a world where the medium truly is the message, and the message is: "Don't take anything too seriously."

RAY (aside or narration):

For our audience, "Now...this" culture refers to how television – and now digital media – presents information. On the evening news, for example, a horrifying report about war or famine might be followed immediately by, say, a story about a celebrity's new hairstyle. The anchor transitions with a smile and the phrase: "And now... this."

There's no time to reflect, to connect, to ask "what does this mean?" Each segment is isolated. Emotional whiplash becomes normal. Everything is flattened into equal weight and delivered with the same tone. A natural disaster and a dog-on-a-skateboard video are just... content. So Neil, now we have a situation where the news is more pervasive than ever, but also more fragmented.

POSTMAN:

Exactly. TV news, and now its digital descendants, prioritize ratings and engagement over truth. The result is a public that's informed in the shallowest sense – bombarded with images and headlines, but lacking context. We've traded depth for immediacy, and substance for spectacle.

RAY:

Neil, in *Amusing Ourselves to Death*, you warned that television would transform not just news, but all public discourse into entertainment. Now, with screens in every pocket and social media as the new public square, how do you see your warnings reflected in 2025?

POSTMAN:

The transformation is even more profound than I imagined. Television was just the beginning. Today, every platform – news, politics, even education – competes for attention by entertaining rather than informing. The "Now...this" style I described has become the default: news stories, TikTok clips, and even political debates are delivered in rapid, disconnected bursts. Substance is sacrificed for spectacle, and the line between news anchor and influencer has all but disappeared.

RAY:

Some argue that more people are "informed" than ever before, thanks to digital access. Is that a positive development?

POSTMAN:

It depends on what we mean by "informed." We have more data, but less understanding. The information-action ratio is dangerously low: we're overwhelmed with facts, but few are relevant or actionable. The result is disinformation – not falsehoods, but contextless, irrelevant information that distracts us from genuine understanding and civic engagement. We hear about disasters across the world, but we can't do anything about them – so we learn to tune out. The more helpless we feel, the more apathetic we become. It's not information overload – it's action underload.

RAY:

Speaking of scrutiny – do we even know where we are anymore? You mentioned tools shape minds. What about, say, Google Maps?

POSTMAN:

You've outsourced your spatial intelligence. The ability to build a mental map of your world – gone. Technology promises utility. But it often replaces old knowledge with dependency. You gain speed and lose understanding.

RAY:

So even maps are part of the problem?

POSTMAN:

Not the tool itself – but our blind trust in it. *Amusing Ourselves* wasn't just about content – it was about surrendering thinking to convenience.

RAY:

Neil, in your book *Technopoly*, you describe a society that has surrendered its culture to technology. Technopoly is your term for a society that not only uses technology, but 'surrenders' to it. In a technopoly, we stop asking whether a tool is good or necessary. We just assume new = better. Every social problem becomes a technical challenge. And human judgment, meanwhile, gets outsourced to machines, metrics, and algorithms.

What does that mean for us in 2025?

POSTMAN:

In a technopoly, technology is not simply a collection of tools; it becomes the defining force of culture itself. We begin to seek our values, our sense of purpose, and even our understanding of what it means to be human, from our machines rather than our traditions or shared stories. Technology stops serving culture and instead becomes the culture.

RAY:

How does this surrender actually happen?

POSTMAN:

It's rarely dramatic. Technology brings what I call "ecological change" – it doesn't just add new tools; it transforms the entire environment. Old ways of knowing – wisdom passed down through stories, rituals, and lived experience – are overshadowed by data, algorithms, and efficiency. The result, which we can see today, is a flattening of culture, where diversity, ambiguity, and complexity are replaced by what can be measured, processed, or optimized.

RAY:

So, what do we lose when we let technology define our culture?

POSTMAN:

We risk losing the depth and richness that make human life meaningful. Our traditions, languages, and unique ways of seeing the world are at risk of being streamlined into a homogenized, technology-approved version of culture. The tie between information and human purpose is severed; we become flooded with information but starved for meaning.

RAY:

Is there a way to resist this surrender?

POSTMAN:

Absolutely. We must become what I call "loving resistance fighters" – people who question the role of technology, who insist that our tools serve our deepest values rather than dictate them. Education should foster critical engagement, not blind acceptance. We must remember that technology is both friend and enemy, and it is up to us to decide which it will be.

RAY:

Now, in 2025, we don't just have television. We have TikTok, Instagram, YouTube, Twitter... sorry, X. Every teen with a phone is an influencer. Every uncle is a conspiracy theorist.

POSTMAN:

I can see that clicks replaced the remote. And screens shrank, but the effect magnified. Instead of one TV in the living room, you now carry your television, your news, your vanity mirror, and your dopamine dispenser in your pocket.

RAY:

And it's always watching us back.

POSTMAN:

Television never asked for your location. TikTok does. Television didn't sell your behavior to data brokers. Instagram does. Your Faustian bargain now includes geotags, surveillance capitalism, and the illusion of "connection."

RAY:

And that's before we even get to the AI stuff.

POSTMAN:

You've taken my concern about visual amusement and added automation, fragmentation, and a business model that thrives on addiction and moral blindness.

RAY:

Let's talk about influencers. They're the new educators. Lifestyle gurus with ring lights. Health tips from people who barely read. It's "edutainment" with a Venmo link.

POSTMAN:

A perfect example of what I called the loss of epistemological order. When we no longer ask who is speaking, what is their authority, and by what logic, we become a culture of vibes and virality.

Now, even our learning is performance. And I use "learning" loosely.

RAY:

What about OnlyFans?

POSTMAN:

Only what?

RAY:

OnlyFans. It's a pay-per-view subscription model where... let's say "boundary-free self-expression" is monetized. Recently scrutinized for hosting underage content.

POSTMAN:

Let me have a quick look at what you're talking about. (pause) Whoa, I see what you mean about 'boundary-free'... This is a grim extension of my concern that ethical trade-offs become invisible in media built for gratification. It's a mirror of our attention economy – where pleasure outpaces principle, and speed outruns scrutiny.

RAY:

Let's talk about kids. You wrote *The Disappearance of Childhood* – what did you mean by that title?

POSTMAN:

Childhood, as a concept, depends on the idea of gradual exposure to adult knowledge. Literacy once created a boundary – children couldn't access adult material until they could read. But television – and now smartphones – obliterate that boundary.

RAY:

So kids today are just small adults with ring lights?

POSTMAN:

Precisely. When media exposes children to adult themes – violence, sexuality, consumerism – without context or guidance, it accelerates their psychological development but stunts their emotional and ethical growth. Childhood becomes a marketing demographic, not a protected stage of life.

RAY:

You argued that television would erase the boundary between childhood and adulthood. Has that come to pass?

POSTMAN:

More than I ever imagined. Childhood, as I wrote, was a social construct born with the printing press – a time when literacy created a gap between what children and adults could know. Television, and now digital media, have erased that gap. Today, children are exposed to adult knowledge, language, and imagery from the moment they can swipe a screen. The mysteries of adulthood – sex, violence, death – are no longer hidden. The result is a blurring of roles: children grow up faster, while adults cling to youth.

RAY:

So, in 2025, is childhood gone?

POSTMAN:

It's certainly vanishing. Social media, streaming, and AI companions mean that children are always "plugged in." When media is accessible to everyone, regardless of age or skill, the distinctions between child and adult dissolve. We see it in fashion, language, even in the way children and adults consume the same content, often side by side. The innocence and gradual maturation that defined childhood for centuries are fading.

RAY:

You also wrote *Teaching as a Subversive Activity*. What did you mean by "subversive"?

POSTMAN:

I meant that education should challenge the status quo. It should teach students to ask questions like: "What's the source of this information?" "What's the bias?" "What's being left out?" But instead, schools became places where students memorize answers to questions they didn't ask.

RAY:

And now we've got AI doing the homework.

POSTMAN:

Which makes the problem worse. If students outsource thinking to machines, they lose the habit of inquiry. Subversive teaching means helping students become skeptics – not cynics, but thoughtful interrogators of their world.

RAY:

So, in Teaching as a Subversive Activity, you and Charles Weingartner called for schools to challenge the status quo. In 2025, with algorithmic instruction, standardized testing, and remote learning, what would you say to teachers?

POSTMAN:

Teaching is more subversive – and more necessary – than ever. The danger today is not just conformity, but passivity. Algorithms personalize content, but they also reinforce biases and discourage questioning. True education must empower students to ask uncomfortable questions, to see their own culture as strange, and to resist the easy answers offered by technology.

RAY:

How can teachers keep learning meaningful in such a digital, data-driven environment?

POSTMAN:

By making inquiry, relevance, and adaptability the heart of their practice. Encourage students to challenge the sources of their information – especially those delivered by screens. Foster environments where dialogue, debate, and critical thinking thrive. The goal is not to produce obedient workers or passive consumers, but thoughtful, skeptical citizens who can navigate – and reshape – the technological society they inherit.

RAY:

What about the role of technology in the classroom?

POSTMAN:

Technology should serve learning, not dictate it. Use digital tools to enhance exploration and creativity, but never let them replace the human elements of teaching: conversation, mentorship, and the shared pursuit of meaning.

RAY:

So, what's your advice for educators, parents, and anyone trying to raise a thoughtful human in 2025?

POSTMAN:

Ask better questions. Teach your children to ask better questions. And stop assuming that more information equals more wisdom. It doesn't. Wisdom requires context, reflection, and restraint. None of which your platforms encourage.

RAY:

Let's shift to organizations. Businesses have gone full TikTok. Marketing. Recruitment. Even Corporate Social Responsibility.

POSTMAN:

Ah yes – corporate virtue-signaling in carousel format.

RAY:

What would you say to management scholars?

POSTMAN:

Stop pretending these tools are neutral. Social platforms reshape what counts as communication inside and outside the organization. They blur the line between brand, employee, citizen, and meme.

Look at the "unmanaged spaces" – where AI writes your ad copy, auto-generates your policy brief, or responds to customer complaints. The message isn't just managed – it's automated. And still nobody's asking: what is the epistemology of this communication?

RAY:

Let's get into finance. Crypto. NFTs. Meme stocks. Dogecoin. Monkey JPEGs selling for $80,000. Decentralized value transfer through blockchain hype. Sound familiar?

POSTMAN:

It's the stock ticker married to the Vegas slot machine, officiated by Reddit.

And again, we confuse information systems with wisdom. Blockchain might be revolutionary. But you've turned it into performance finance. You didn't just amuse yourselves – you commodified that amusement into fake economics.

RAY:

And now, AI is writing news articles, essays, and podcasts like this one.

POSTMAN:

Machines are now imitating cognition. But can they reason? Feel? Take responsibility? If you outsource judgment to machines, you're not innovating – you're institutionalizing what Enlightenment thinkers called "organized immaturity." You no longer think for yourselves because it's easier to let the machine think for you.

RAY:

In the current craziness that is global politics, I can't let you go without talking about today's polarized politics, and of course, the phenomenon that is Donald Trump.

POSTMAN:

Ah, yes. Donald Trump. The inevitable result of a culture that mistook a television host for a statesman. Let me be blunt – your political discourse didn't just fall apart overnight. It was replaced, systematically, by show business.

I tried to warn you in the 1980s. I said, when politics becomes entertainment, the citizen becomes an audience. And audiences don't deliberate. They applaud. They boo. They vote for the character with the best one-liners and the most dramatic entrance. You didn't elect a president – you cast a leading man.

Seeing that the 45th (and 47th – hang on, are you telling me they voted him back in?) President of the United States was a reality television star whose greatest qualification for office was knowing how to dominate a news cycle, I feel no surprise. Mild heartburn, perhaps. But no surprise.

This wasn't an aberration. It was the logical conclusion of a process I documented in the 1980s – the transformation of public discourse into entertainment. It's not that Donald Trump caused the degradation of politics. He simply understood how to surf it better than anyone else.

When I wrote *Amusing Ourselves to Death*, I warned that television – your dominant medium at the time – rewards only what is visually stimulating, emotionally simple, and instantly gratifying. It is a medium that hates complexity. And democracy, dear listeners, is nothing but complexity. It requires context, argument, history, delayed gratification, and the ability to tolerate ambiguity.

RAY:

What about the polarization that has happened in politics today? This is something that started long before Trump came on the scene.

POSTMAN:

Polarization? That's just another word for narrative simplicity. You're no longer arguing ideas. You're choosing sides in a morality play. Red hats versus blue hashtags. The nuance has been edited out, like bad ratings.

And here's the kicker: the medium trained you to like it that way. Twitter didn't cause polarization – it rewarded it. Cable news didn't create extremism – it monetized it. And TikTok? That's just politics on amphetamines.

Your democracy, I'm afraid, has become a game show. And the grand prize? Your attention. Which, if I may say, you've given away far too cheaply.

You've confused spectacle with substance, drama with dialogue. And now you wonder why everyone's shouting and no one's listening. But don't worry – it makes great television.

Polarization is not just a side effect of ideological conflict. It is, in your era, a design feature. Every modern communication platform you use – from Facebook to YouTube to TikTok – is built on algorithms that reward outrage, amplify tribalism, and curate the world to confirm your biases. That's not a conspiracy. That's just good business. Anger gets clicks. Division holds attention. Conflict keeps you scrolling.

You are, quite literally, being fed the version of reality most likely to keep you emotionally agitated – and therefore engaged. That's not democratic discourse. That's operant conditioning.

And don't get me started on debates. In my day, a political debate was an argument over policy. In your day, it's an exchange of insults framed by dramatic camera angles, with moderators trained more in timekeeping than in Socratic inquiry. It's not that your leaders don't speak in paragraphs anymore. It's that paragraphs don't fit on TikTok. If Lincoln had to debate Douglas in 2025, the winner would be the one who could turn a log cabin into a viral meme.

Trump is not unique. He is what I would call a symbolic endpoint. A kind of media-virus – perfectly adapted to a host culture where every issue must be reduced to a slogan, every policy a tweet, every opponent a villain. His power lies not in argument, but in performance. He understands what the medium demands: attention, spectacle, and simplicity. His opponents try to beat him with facts, but that's like bringing a library to a gunfight.

You see, you've been taught – by your media environments – that truth is not determined by logic or evidence, but by repeatability and visibility. Whoever says it loudest, most often, and with the best production values – wins.

RAY:

So, what happens to a society where the very form of its communication undermines its content?

POSTMAN:

You get a public who doesn't know the difference between a policy and a personality. You get news that feels more like professional wrestling. You get "debates" that are edited for drama rather than depth. And eventually, you get a citizenry that stops acting like a deliberative body – and starts behaving like a fan base.

That's not democracy. That's an audience. And what do audiences do? They cheer. They boo. They wait for the next season.

You've turned the republic into reality television. And I can tell you from beyond the grave: the ratings may be high, but the plot is headed somewhere dark. If you want to fix it, don't just change the channel. Change the medium.

RAY:

Neil Postman, thank you for returning to this beautiful nightmare. Any last words for our 2025 audience?

POSTMAN:

Read slowly. Think deeply. Speak with intention. And stop calling everything "content."

Reflection and Commentary: Education, Not Technophobia

Postman's central insight was that *the medium matters*. It does not merely deliver content neutrally; it shapes how content is received and what kinds of content can even exist. His famous comparison of Orwell and Huxley remains a cultural touchstone: Orwell feared oppression; Huxley feared distraction. Postman believed Huxley had the better forecast.

And in many ways, the 2020s have vindicated him. From TikTok to Twitter, we now inhabit a media ecosystem optimized for emotion, brevity, and engagement – not truth, nuance, or public deliberation. Serious discussion drowns in the noise. Complex issues are flattened into memes. Education is gamified. Childhood is branded.

Postman didn't hate technology. But he demanded that we ask harder questions about its cultural impact. Questions we still avoid today:

- What happens when public discourse is dominated by images and not ideas?
- What values are encoded in our platforms?
- Are we educating citizens, or training consumers?

He also understood that technology isn't just external. It colonizes our inner life. The way we think, the way we speak, even the metaphors we use to understand the world are shaped by the tools we use.

In *Technopoly,* Postman warned of a society that surrenders its culture to the sovereignty of technology. Not a totalitarian regime. Not a dystopia with secret police. But a society that forgets what it stands for, because it becomes enchanted by what it can do.

Sound familiar?

His solution wasn't technophobia. It was education. Real education. Teaching people not just to use tools, but to critique them. To see technology not as fate, but as choice.

As the fictional Postman says in our imagined interview: "You have confused access with understanding. Connection with meaning. And you have turned learning into a game of points and prizes."

That line could have been written yesterday.

Further Reading and Exploration

Primary Works
Postman, N. (1985). *Amusing Ourselves to Death: Public Discourse in the Age of Show Business.* Viking.
Postman, N. (1992). *Technopoly: The Surrender of Culture to Technology.* Vintage.

Postman, N. (1994). *The Disappearance of Childhood.* Vintage.
Postman, N. & Weingartner, C. (1969). *Teaching as a Subversive Activity.* Dell.

Modern Commentary and Biographical Sources
Levinson, P. (1999). *Digital McLuhan: A Guide to the Information Millennium.* Routledge.
Vaidhyanathan, S. (2018). *Antisocial Media: How Facebook Disconnects Us and Undermines Democracy.* Oxford University Press.
Turkle, S. (2011). *Alone Together: Why We Expect More from Technology and Less from Each Other.* Basic Books.

Adjacent Readings
Carr, N. (2010). *The Shallows: What the Internet Is Doing to Our Brains.* W. W. Norton & Company.
Lanier, J. (2018). *Ten Arguments for Deleting Your Social Media Accounts Right Now.* Henry Holt & Co.

2. Marshall McLuhan – The Medium Prophet

"There is absolutely no inevitability as long as there is a willingness to contemplate what is happening."

Preface: The Medium That Swallowed the Message

Marshall McLuhan was not a technologist. He was a literary scholar turned media theorist who saw what few others could: that the form of communication shapes human thought more profoundly than its content. Born in Edmonton, Canada, in 1911, McLuhan became a global intellectual icon in the 1960s. His books *The Gutenberg Galaxy* and *Understanding Media* helped invent an entirely new way of thinking about how we relate to media and technology.

McLuhan famously claimed, "The medium is the message" – a phrase that confused, annoyed, and inspired a generation. What he meant was deceptively simple: a television isn't just a conduit for content like news or entertainment; it restructures the way we process reality. A book creates a different kind of mind than a radio, and a smartphone yet another.

In our era of constant screen-swiping, ambient data collection, and hyper-connected distraction, McLuhan's cryptic phrases now seem like warnings from a time traveler. He predicted global connectivity before the internet, memes before social media, and media echo chambers long before we had names for them.

He also had a wicked sense of humor and a gift for provocation. It's no surprise he made a cameo in *Annie Hall*, popping up from behind a theater display to tell an academic blowhard: "You know nothing of my work."

So, we brought him back. To let him have another word. Or maybe several thousand.

Fictional Interview: Marshall McLuhan in 2025

RAY:

What if I told you the most famous media theorist of the 20th century wasn't some Silicon Valley futurist or venture-funded disruptor – but a man in a tweed suit who sounded like he just wandered out of a Renaissance fair?

Yep. We brought back Marshall McLuhan.

Marshall McLuhan didn't just write about media. He was media. A Canadian professor and Catholic mystic in tweed, he became an unlikely pop culture figure in the 1960s, throwing intellectual firecrackers into the worlds of journalism, advertising, and academia. He said things like "the medium is the message," "the global village," and "the medium is the massage," and somehow got people to listen. He was a serious thinker with a gift for cryptic phrasing, and he predicted the coming digital age in eerie, riddling bursts.

But ask the average person about him today, and they won't mention Understanding Media. They'll mention *Annie Hall*.

McLuhan's cameo in the 1977 Woody Allen film is iconic. A man in line at the movies pontificates loudly about media theory. Allen, annoyed, pulls McLuhan from behind a sign to silence the man with the immortal line: "You know nothing of my work." It was perfect. And it forever made McLuhan, in his words, "a punchline with tenure."

That's him. That's McLuhan. That cameo made him a meme before memes existed. But behind the punchline was a prophet.

Today, we raise him from the dead – because, well... they tried to warn us. Marshall, welcome back. Still annoyed about Annie Hall?

MCLUHAN (vintage Canadian lilt):

Not annoyed. Just resigned. My entire career reduced to one snarky movie line. Typical.

RAY:

Let's go deeper. Who were you?

MCLUHAN:

An English literature professor turned media mystic. I didn't study content – I studied form. I said, "The medium is the message." People thought I was being clever. I was being blunt. The platform matters more than the content it carries.

Television rewired your brain. The internet reprogrammed it. And social media? That's a mood disorder in app form.

RAY:

You coined the phrase "the global village" to describe the effect of electronic media collapsing time and space. You envisioned a world where everyone was connected in real-time, participating in shared experience.

The phrase global village sounded quaint and cute in the '60s.

MCLUHAN:

It wasn't meant to be cute. I said electronic media would collapse time and space – connect everyone in real time. I didn't say they'd like it.

RAY:

Because what we got was less Sesame Street, more Thunderdome.

MCLUHAN:

Yes. What I didn't predict was how much the global village would resemble an unruly town square: tribal, emotional, mob prone. In 2025, every major platform – Facebook, Reddit, Twitter (sorry, X) – acts less like a forum and more like a coliseum. Outrage is rewarded. Consensus is rare. Gossip spreads faster than facts. You think you're connected, but you're actually fragmented – each person in their own echo chamber, shouting.

RAY:

As you said back in the 1970s, "We shape our tools and thereafter our tools shape us." You warned that we'd mistake connection for community. Well, you weren't wrong. You also introduced the idea of "hot" and "cool" media. Hot media are data-rich and low in participation – like film, print, or radio. Cool media, meanwhile, are low-definition, requiring audience engagement – like conversation or cartoons. With that in mind, what do you make of something like, say, TikTok?

MCLUHAN:

It actually breaks my framework. It's "hot" in terms of sensory overload – images, sound, text – but "cool" in attention span. You don't watch; you scroll. You don't absorb; you swipe. You perform. You are performed.

RAY:

So... lukewarm?

MCLUHAN:

Lukewarm and numbing. The content is secondary. The real impact is what it trains you to become. It's ultra-hot in data (images, music, subtitles, filters) but emotionally numb. It invites constant interaction while numbing attention. It blurs the line between performer and audience. It trains the viewer to swipe before they feel. To me, this isn't just a new platform; it's a new environment for consciousness. In 2025, we're not just extending ourselves. We're replacing parts. Smartphones have become prosthetic brains.

RAY:

You wrote that technology extends the senses. The wheel extends the foot. The phone extends the voice. But in 2025...

MCLUHAN:

...you're not just extending yourself. You're outsourcing yourself.

Your phone isn't a tool anymore. It's a prosthetic memory. Your AI isn't a servant – it's a second self. AI is replacing writers, artists, and even therapists. Identity is built in pixels and validated by metrics. I once quipped, "First we build the tools, then they build us." Now, I would say, "Then they own us." And now you're building tools that imitate you... poorly.

RAY:

Like chatbots?

MCLUHAN:

Like me, apparently.

RAY:

What do you make of generative AI?

MCLUHAN:

Machines that speak without intention? That's not communication. That's ventriloquism. AI isn't a medium – it's a simulation of a medium. It anesthetizes thought. It confuses fluency with meaning.

RAY:

But we get so much content!

MCLUHAN:

Yes, and so little conversation. You used to debate Plato in line at the theater. Now you argue with a bot named Chad-47 on Discord. Do you call that progress?

RAY:

Let's talk about media environments. You said media aren't just tools – they're environments. What does that mean in 2025?

MCLUHAN:

A fish doesn't know it's in water until it's out of it. Likewise, you don't notice your media until they change. TikTok, Instagram, AI chatbots – these are not just platforms. They are environments that reconfigure your senses, your attention, your time.

RAY:

So, we're not just using media – we're living inside them?

MCLUHAN:

Precisely. You no longer go online. You are online. Your nervous system is plugged into a perpetual feed. The environment is invisible because it is total.

RAY:

You also developed something called the "tetrad" – a four-step process to analyze media. Can we apply that to, say, TikTok?

MCLUHAN:

Let's try. The tetrad involves asking four questions about a medium – what it enhances, what it makes obsolete, what it retrieves and what it reverses into when it's pushed to its extremes. So, let's look at TikTok.

- What does TikTok enhance? Instant expression. Micro-celebrity. The illusion of intimacy.

- What does it make obsolete? Reflection. Patience. The long-form argument.

- What does it retrieve? Oral culture. Performance. The village square – now digitized.

- What does it reverse into when pushed to extremes? Surveillance. Narcissism. A feedback loop of distraction.

RAY:

So even a dance app is a cultural transformer?

MCLUHAN:

There are no "just" media. Every medium is an architecture of perception. And architecture always has consequences.

RAY:

Is it true that you said you predicted the internet?

MCLUHAN:

I never said that! I predicted connection. But I didn't predict your uncle radicalizing himself via Facebook memes.

RAY:

Any final thoughts for our 2025 audience?

MCLUHAN:

I never hated technology. I hated unexamined environments. You need to step outside your media to ask: What is this doing to me? What does it let me say – and what does it silence? I wrote when I was alive that the medium is the message. But today, the medium is the self.

You don't just use media. You are media. You're shaped by it. You're performed through it. You are the message, whether you like it or not.

Reflection and Commentary: The Return of the Tetrad

McLuhan's brilliance wasn't in predicting particular technologies, but in giving us a grammar to understand their effects. His four-part "tetrad" – what a medium enhances, obsolesces, retrieves, and reverses into – remains a powerful diagnostic tool.

Take the smartphone. It enhances communication, obsolesces landlines and attention spans, retrieves the oral culture of constant chatter, and reverses into isolation and surveillance. That's a McLuhan analysis.

He argued that media are not just tools; they are environments. Invisible environments, like water to a fish. And if you don't notice your environment, you can't critique it. That's why McLuhan often sounded more like a Zen master than a professor: he was trying to shock us into awareness.

In 2025, we swim in an ocean of media forms that flatten nuance, encourage tribalism, and rewire our cognition. As McLuhan might say, we've extended our nervous system into the ether – and now everything twitches.

He didn't claim technology was evil. But he insisted that we become *conscious* of it. His greatest insight? That we shape our tools, and thereafter they shape us. He warned: "First we build the tools, then they build us." Maybe now, we'll listen.

Further Reading and Exploration

Primary Works
McLuhan, M. (1964). *Understanding Media: The Extensions of Man*. McGraw-Hill.
McLuhan, M. & Fiore, Q. (1967). *The Medium is the Massage: An Inventory of Effects*. Penguin.
McLuhan, M. (1962). *The Gutenberg Galaxy: The Making of Typographic Man*. University of Toronto Press.

Modern Commentary and Biographical Sources
Cavell, R. (2002). *McLuhan in Space: A cultural geography*. University of Toronto Press.
Federman, M. (2004). "What is the Meaning of The Medium is the Message?" University of Toronto (online article).
Levinson, P. (1999). *Digital McLuhan: A guide to the information millennium*. Routledge.
Marchand, P. (1989). *Marshall McLuhan: The Medium and the Messenger*. MIT Press.

Strate, L. (2008). *Echoes and Reflections: On Media Ecology as a Field of Study*. Hampton Press.

Adjacent Readings
Ong, W. J. (1982). *Orality and Literacy: The Technologizing of the Word*. Methuen.
Postman, N. (1985). *Amusing Ourselves to Death: Public Discourse in the Age of Show Business*. Viking.

3: Guy Debord – The Society of the Spectacle

"You are not conquered. You are convinced."

Preface: The Spectacle Takes the Stage

Guy Debord was a revolutionary who didn't seek power, a filmmaker who disdained narrative, and a theorist who saw through the curtain of modern life. Born in Paris in 1931, Debord became a leading voice of the Situationist International, a radical leftist collective that blended Marxism, art, and existential critique. In 1967, he published *The Society of the Spectacle*, a work that would echo through the uprisings of May 1968 and well beyond.

His central thesis? That in late capitalism, all lived experience had been replaced by representation. The real was now mediated through images, and the image had become the dominant mode of social control. Not simply propaganda or advertising, but a deep, structural substitution of life itself. As he wrote: "All that once was directly lived has become mere representation."

Debord rejected stardom. He avoided interviews. He made films few could bear to sit through – full of aphorisms, static, and repetition. But he understood the danger of a culture that no longer distinguishes between performance and reality. In the age before smartphones, he warned that spectacle would become the organizing principle of society.

Now, in 2025, his vision reads not as prophecy, but as operating manual. From livestreamed identities to branded activism, from TikTok revolutions to monetized rebellion, the spectacle has metastasized. Debord doesn't just feel relevant. He feels like the ghost in the machine.

And so, we summoned him, to see what he would say.

Fictional Interview: Guy Debord in 2025

RAY:

Guy Debord. French. Marxist. Filmmaker. Revolutionary. Founder of the Situationist International – a group of radical artists and thinkers who believed that modern life had been hijacked by appearances.

In 1967, Debord published a book that would become a bible for late-century rebellion: *The Society of the Spectacle*. In it, he argued that in modern capitalism, everything authentic – labor, thought, relationships, identity – had been replaced by images. Not just advertisements, but life itself had become a performance.

We no longer live directly – we represent our lives. Consume curated versions of others. Pose. Perform. Scroll. Sell.

Sound familiar?

He wrote: "All that once was directly lived has become mere representation."

Debord saw not just the rise of media – but the rise of mediation. And it terrified him. He lived in defiance of it. Refused interviews. Burned bridges. Shot abstract films with no characters. He drank heavily, wrote furiously, and died by suicide in 1994 – just before the internet fully absorbed us.

But now, in 2025, the spectacle is not a theory. It is the operating system. Reality is staged. Truth is filtered. And life – already commodified – has become content.

When we brought him back, Debord took one look at a TikTok influencer livestreaming a staged protest and muttered: "It's worse than I imagined."

Mr. Debord – welcome to the future. You warned that capitalism would evolve not just to sell products, but to sell experiences, identities, desires. Now, we have Instagram "aesthetics," monetized trauma, and AI-generated nostalgia. Did the spectacle win?

DEBORD (calm, dry):

The spectacle does not win. It replaces victory with visibility. You are not conquered. You are convinced.

RAY:

You described the spectacle as a "social relationship mediated by images." In 2025, we communicate almost entirely through screens – text, emojis, reaction gifs, stories. Even activism is now branding. How would you describe the modern condition?

DEBORD:

You are alienated from your labor. Now, you are also alienated from your persona. You do not express yourself. You manage your image. You consume your reflection – and confuse it for a self.

RAY:

You were part of the Situationist movement – people who believed in "constructing situations," disrupting the normal flow of daily life to awaken people from their passive roles. But now, disruption is a brand. Chaos is aesthetic. Even rebellion is merchandised. Is detournement still possible?

DEBORD (grim smile):

Only if it resists simulation. Today's rebellion is often performed – for clicks, for sponsorships. But true detournement is uncomfortable. It does not flatter. It exposes. You must interrupt the spectacle without becoming it.

RAY:

Let's talk about algorithms. Today, what you see is curated by systems designed to predict – and manipulate – your attention. Did you anticipate this level of behavioral control?

DEBORD:

Yes. But I underestimated your surrender. You call it personalization. It is pre-emption. You are no longer choosing. You are being chosen for – by a logic you cannot see.

Even your outrage is algorithmic. Your rebellion has been budgeted for.

RAY:

You once wrote, "The spectacle is the sun that never sets over the empire of modern passivity." That was before 24/7 news. Before social media. Before people began livestreaming their lives, deaths, crimes. How would you describe the current spectacle?

DEBORD:

It has achieved omnipresence. In my time, it colonized your labor and leisure. Now, it colonizes your dreams, your intimacy, your grief. The spectacle no longer needs to convince. It only needs to distract. You forget what you were feeling before the next clip loads.

RAY:

But some say we're more informed than ever. We can see through propaganda. Isn't transparency a kind of liberation?

DEBORD (shakes head):

No. You are drowning in clarity. And yet you see nothing. The spectacle does not hide reality. It floods it. When every truth is immediately juxtaposed with parody, doubt, or simulation – truth becomes optional.

RAY:

Your ideas influenced the May '68 uprisings in France. Students and workers quoting you in graffiti: "Live without dead time." "Beneath the pavement – the beach." What would you say to young people today, stuck in cycles of debt, burnout, and curated unreality?

DEBORD:

Reclaim time. Do not measure it in productivity. Do not monetize it through visibility. Wander. Waste. Drift. The most radical act in a society of control is to be useless to it.

RAY:

But isn't that dangerous? To opt out, unplug, resist visibility – don't people risk obscurity? Irrelevance?

DEBORD:

Yes. But irrelevance is not failure. To disappear from the spectacle is not to vanish – it is to reappear in life. Obscurity is where the authentic hides. In friendship without selfies. In thought unshared. In joy untagged.

RAY:

You filmed works that rejected narrative, logic, even clarity. You layered aphorisms over disjointed imagery. Some said you were trying to make cinema impossible. What role does art have now, in a world of infinite content?

DEBORD:

Art must wound the illusion. If it entertains, it must betray. If it comforts, it must confess. If it is beautiful, it must ask why. The artist's role is not to join the spectacle – but to derail it.

RAY:

Final question: You ended your life in 1994. Before smartphones. Before TikTok. Before deepfakes, virtual influencers, AI girlfriends. If you had lived to see all this – what would you have done?

DEBORD (quietly):

I would have written less, and listened more. To the wind. To silence. To unrecorded laughter. Then, perhaps, I would have said: "Leave the screen. Follow the shadow. Find the unfilmed life."

RAY:

Guy Debord warned us the real would be replaced by the represented. That we would forget how to live – because we were too busy watching others pretend. He saw the screen not as window, but as veil. Not as connection, but as cage.

He didn't ask us to unplug. He asked us to wake up. To taste again. To touch again. To revolt – not on camera, but in the street. Not in hashtags, but in presence.

Reflection and Commentary: We Are the Spectacle

What makes Debord chillingly prescient is that his critique was never about media content. It was structural. In his view, capitalism's final stage was not production or even consumption – but the mediation of all reality. The image becomes more real than the real. And worse: the citizen becomes a spectator.

In the era of social media, influencer culture, algorithmic feeds, and clickbait outrage, Debord's words hit with new force. We don't just consume images. We curate ourselves as images. Our outrage is monetized. Our dissent is sold back to us in tote bags and branded livestreams.

Where McLuhan argued that "the medium is the message," Debord pushed further: the medium becomes reality. It replaces it. And this new reality is passively accepted because it flatters, distracts, and entertains.

Debord's insistence on detournement – turning spectacle against itself – now feels quaint in a world where rebellion is aestheticized in real time. A protest sign might trend before the protest begins. Outrage becomes brand strategy. Debord would argue we have lost even our illusions of freedom.

And yet, he offers no easy solution. Instead, he offers a challenge: wake up. Reclaim time. Seek the unspectacular. As he tells Ray in the interview: "To disappear from the spectacle is not to vanish – it is to reappear in life."

That's not nihilism. It's resistance.

Further Reading and Exploration

Primary Works
Debord, G. (1967). *The Society of the Spectacle*. Buchet-Chastel/Zone Books (1994 English edition).
Debord, G. (1988). *Comments on the Society of the Spectacle*. Verso.

Modern Commentary and Biographical Sources
Jappe, A. (1999). *Guy Debord*. University of California Press.
Hussey, A. (2001). *The Game of War: The Life and Death of Guy Debord*. Pimlico.
Sadler, S. (1998). *The Situationist City*. MIT Press.

Adjacent Readings
Baudrillard, J. (1981). *Simulacra and Simulation*. Éditions Galilée / University of Michigan Press.
Sontag, S. (1977). *On Photography*. Farrar, Straus and Giroux.
Wark, M. (2011). *The Beach Beneath the Street: The Everyday Life and Glorious Times of the Situationist International*. Verso.

4: David Foster Wallace: The Cost of Consciousness

"The most radical act in a culture of distraction is to sit still long enough to notice you're still here."

Preface: The Novelist of Our Neural Networks

David Foster Wallace (1962–2008) was not a futurist, technologist, or academic theorist. Yet, few writers captured the emotional and existential toll of modern life as presciently as he did. With a mind equal parts linguistic gymnast and moral philosopher, Wallace wrote about addiction, irony, attention, loneliness, and the quiet, often unbearable burden of being awake to the world.

His sprawling 1996 novel *Infinite Jest* imagined a society paralyzed by its own entertainment, long before anyone had a smartphone in their pocket. His nonfiction work, especially the essays collected in *A Supposedly Fun Thing I'll Never Do Again* and *Consider the Lobster*, probed American culture with humor and precision. And his 2005 commencement address, "This Is Water," remains a singular meditation on awareness, compassion, and choice.

Wallace didn't write about AI, social media, or algorithmic addiction directly. He didn't have to. He wrote about their psychological ancestors: the internal loops of self-consciousness, the performative ego, and the hollowness of curated experience.

In resurrecting him for this series, we aren't indulging in nostalgia – we're acknowledging that Wallace may have been the first writer of the post-internet condition. And we didn't even notice until it was too late.

Fictional Interview: David Foster Wallace in 2025

RAY:

Today's guest didn't give us theories or manifestos. He gave us mirrors. David Foster Wallace was a novelist, essayist, and reluctant prophet of the digital age.

Born in 1962, raised on philosophy and tennis, Wallace came of age at the precise moment when American life turned into entertainment – and he never got over it. His landmark 1996 novel *Infinite Jest* dissected a world addicted to pleasure, performance, and the slow death of attention. It wasn't dystopian – it was too familiar for that.

He also wrote essays on everything from cruise ships to talk radio to grammar, but they all shared a deep theme: the terror and beauty of consciousness in a world that won't let you sit still.

Wallace understood that irony can curdle into paralysis. That freedom can feel like drowning. That we are, all of us, just one distraction away from noticing how alone we are.

In 2005, in what may be the greatest commencement address ever given, he told a room full of college graduates:

"The really important kind of freedom involves attention, and awareness, and discipline… and the ability to truly care about other people."

He died three years later by suicide, after a lifelong struggle with depression.

In 2025, we brought him back. He opened Instagram. Watched a sponsored ad for "mindfulness gummies." Then said: "You turned alienation into an industry."

David, welcome to the future. In *Infinite Jest*, you imagined a piece of entertainment so addictive it destroyed anyone who watched it. That idea felt speculative in 1996. How does it feel now?

WALLACE:

Honestly? Pretty tame. What I didn't quite grasp – what I maybe couldn't – was that the most powerful entertainment wouldn't look like a weapon. It wouldn't hypnotize you. It would simply fit. It would flatter you. Adapt to you.

And that's what the feed is. It's not something you fall into once. It's something you live inside of. It becomes the default state. And the real danger isn't that you disappear – it's that you slowly stop noticing you're gone.

RAY:

You wrote about addiction, not just to substances but to distraction itself. In 2025, we're addicted to attention – giving it, getting it, selling it. What do you think that's doing to us?

WALLACE:

It's numbing the very thing that makes us human – our capacity for deep, painful, ecstatic presence. We've been trained – quite literally, operantly conditioned – to respond more than reflect, to curate more than feel.

I don't think people are weak. I think they're exhausted. Because it takes so much more energy to sit with yourself than to escape yourself. And the culture offers you a million ways to escape.

RAY:

You were also critical of irony – what you called "hip detachment." You said it was useful for poking holes in lies, but terrible at building anything new. How does irony function in today's media world?

WALLACE:

Irony has metastasized. It used to be a scalpel. Now it's the wallpaper.

Look – irony is seductive because it lets you be smart without being vulnerable. It lets you mock everything, so you never have to believe in anything. But that kind of detachment, when it becomes constant, is just cowardice in cool clothes.

The real work is belief. It's sincerity. It's risking looking stupid. And in a culture where every sincere thing can be clipped, mocked, memed – sincerity becomes this act of bravery.

RAY:

In your essays, you talked about "default settings" – the unconscious assumptions we live inside. You said real freedom was choosing how to see the world. In a time when our feeds choose for us, is that freedom still possible?

WALLACE:

It's harder now. The "default setting" is literally engineered. You're nudged, primed, targeted, profiled – and rewarded for staying inside the loop.

But yeah, I still think freedom's possible. Because awareness is still possible. And once you notice what's happening – not just intellectually, but viscerally – you start to get some leverage. You start to ask, "Wait – what am I not paying attention to while I'm inside this scroll?" That question is the beginning of freedom.

RAY:

Your work also explored loneliness. Not just physical isolation, but existential loneliness – the kind that comes from feeling like no one else is really there. Has digital life made that better or worse?

WALLACE:

Both. And that's the trap. You can "connect" 24/7. You can feel seen by strangers. But the more you connect through a medium, the more you risk turning into a version of yourself – something optimized, marketable, legible.

Real connection is messy. It's boring. It's embarrassing. It's inefficient. And all your tools – your apps, your filters, your platforms – are designed to scrub that mess out.

RAY:

Let's talk about the self. You once described the ego as this constant voice saying "me, me, me." Today, we're all curating identities online. Performing for audiences. What happens to the self in this environment?

WALLACE:

It gets hungry. And hollow.

See, the thing about performance is that it never ends. Once your sense of identity is tied to external feedback, you're trapped in an arms race with your own image. And that's exhausting. Eventually, you don't even know if you're living your life – or just running the account.

RAY:

You also talked about boredom – not as failure, but as a gateway to attention. But now, boredom is algorithmically eliminated. What are we losing?

WALLACE:

We're losing the room in which your actual self might walk in. Boredom is where your mind gets a chance to wander – to process, to mourn, to remember, to wonder. When that's gone, you're just reacting. You're not choosing. You're being scrolled through.

RAY:

Last question. Someone listening to this is probably doing three other things at once. They're busy. They're tired. They feel like they're falling behind. What would you want to say to them?

WALLACE (gently):

You're not behind. You're alive. And that's already absurd and miraculous.

You don't have to fix the whole system today. You don't even have to log off forever. But maybe – just maybe – you could try sitting still.

Not forever. Just long enough to notice that you're still here. That might be the beginning of something.

RAY:

David Foster Wallace warned us – not in slogans, but in syntax. Not through certainty, but through the hard, trembling work of asking what it means to be awake.

He didn't reject technology. He rejected numbness. And he showed us that the real danger isn't distraction. It's forgetting what we were trying to escape from in the first place.

Reflection and Commentary: Irony, Attention, and the Self in the Age of Infinite Feeds

David Foster Wallace saw the irony-soaked, hyper-mediated future with unnerving clarity. What he could not have predicted was how quickly it would arrive – and how fully we would submit to it.

The central metaphor of *Infinite Jest* was a piece of media so pleasurable it destroyed the viewer's ability to care about anything else. In 1996, it was speculative satire. Today, it's almost quaint. The entertainment he feared wouldn't be a drug, but a mirror: flattering, endless, addictive. Wallace understood that the more media personalized itself to our preferences, the more it would sedate the deeper parts of us. He anticipated that our desire to escape discomfort would mutate into an economy – a system where distraction is monetized, and attention is auctioned.

He also saw irony not as a fun flourish but as a cultural toxin. When sincerity becomes suspect and detachment becomes currency, belief itself is devalued. Wallace worried that a generation raised on sarcasm and posturing would become emotionally illiterate – unable to risk sincerity, intimacy, or even vulnerability. Social media didn't invent this problem; it just industrialized it.

More than anything, Wallace wrote about attention. Not the "grab-it-in-three-seconds" kind, but the kind that requires discipline. Stillness. The willingness to remain present with boredom, with ambiguity, with pain. In his vision, true freedom wasn't doing whatever you want – it was choosing how to think, moment by moment. That kind of freedom is more radical today than ever, because everything in our digital world is designed to automate our thinking and outsource our desires.

What Wallace reminds us is that the enemy is not technology. It's unconsciousness. It's the default setting. It's letting the feed think for you. His voice urges us to reclaim our agency, not through grand revolutions, but through the humble, daily work of paying attention. Noticing. Sitting still.

In the end, Wallace didn't give us easy answers. He gave us better questions. And in a time when everyone has a platform, but few have presence, his is the voice that still whispers: "You get to decide what has meaning."

Further Reading and Exploration

Primary Works
Wallace, D. F. (1996). *Infinite Jest*. Little, Brown and Company.
Wallace, D. F. (2009). *This Is Water: Some Thoughts, Delivered on a Significant Occasion, about Living a Compassionate Life*. Little, Brown
Wallace, D. F. (1997). *A Supposedly Fun Thing I'll Never Do Again*. Back Bay Books.

Modern Commentary and Biographical Sources
Boswell, M. (2003). *Understanding David Foster Wallace*. University of South Carolina Press.
Franzen, J. (2011). "Farther Away" in *The New Yorker* (April 11) (memoir and reflection on Wallace).
Max, D. T. (2012). *Every Love Story Is a Ghost Story: A Life of David Foster Wallace*. Viking.

Adjacent Readings
Franzen, J. (2001). *The Corrections*. Farrar, Straus and Giroux.
Powers, R. (2006). *The Echo Maker*. Farrar, Straus and Giroux.

5: Philip K. Dick: The Simulation Has Crashed

"You are not your machine. Not your data trail. Not your feed. You are what resists."

Preface: The Prophet of Paranoia and Perception

Philip K. Dick (1928–1982) was not merely a science fiction writer. He was a metaphysical detective, a philosopher disguised as a pulp novelist, and one of the most uncannily accurate chroniclers of our fractured digital now. Long before "surveillance capitalism" or "the simulation hypothesis" became cultural shorthand, Dick was already obsessed with the collapse of reality and the erosion of human authenticity.

His oeuvre includes over 40 novels and hundreds of short stories, many of which have been adapted into films: *Blade Runner*, *Minority Report*, *Total Recall*, *A Scanner Darkly*, and *The Man in the High Castle*, to name a few. But to call him prophetic is to understate his weird clairvoyance. Dick didn't so much foresee the future as feel it – through anxiety, hallucination, and relentless interrogation of the self.

He saw that the future wouldn't come wrapped in chrome. It would come wrapped in code. And when it arrived, we wouldn't resist it. We'd subscribe.

Fictional Interview: Philip K. Dick in 2025

RAY:

Today's guest didn't write science fiction to predict the future. He wrote to survive the present. Philip K. Dick was a paranoid, brilliant, deeply human chronicler of reality's collapse. He lived half his life under surveillance – sometimes real, sometimes imagined. He questioned everything: the government, the self, the nature of time, the reliability of memory, even the fabric of reality itself.

He didn't just ask "what if?" He asked: "What is real? What is human? And who the hell is running this thing?"

In more than 40 books, many of which were turned into films and television shows, including Minority Report, Total Recall, *Ubik*, *The Man in the High Castle*, and *Do Androids Dream of Electric Sheep?* (which was turned into the movie Bladerunner) he peeled back the shiny veneer of the American dream – and found layers of simulation, surveillance, and spiritual vertigo.

He didn't just ask what machines might become. He asked: what's left of the human when everything else has been faked?

In 2025, we brought him back. He saw the VR headsets. The AI-generated influencers. The conspiracy-pilled content creators. He rubbed his eyes and said: "You've made the world I hallucinated. Congratulations. It's worse than I thought. But I'm not surprised."

Phil – welcome to the future. You wrote about precogs and replicants, corporate empires and collapsing timelines. But now we have AI therapists, personalized propaganda, and memory feeds that remind us who we are. What do you see?

PHILIP K. DICK:

I see a world where everyone doubts what's real – but no one asks why. You've outsourced perception to machines. You believe your phones more than your senses. You scroll past lies and call it literacy. And you still think you're awake.

RAY:

You wrote obsessively about false realities – entire worlds built to deceive. Do you think we're living in one now?

DICK:

You're not in a simulation. You're in a consensus. A collective illusion held together by screens, habits, and algorithms that learn just enough about you to keep the mask from slipping.

It's not that the world is fake. It's that your relationship to it is mediated by a thousand filters. And somewhere along the line, you stopped asking: who benefits?

RAY:

In *Do Androids Dream of Electric Sheep?*, empathy is the defining human trait. That's how we tell real from artificial. But today, even our emotions are tracked, optimized, predicted. What happens when machines mimic empathy?

DICK:

You get simulation without sincerity. You get chatbots trained to comfort you – so you don't bother your friends. You get grief apps that analyze your voice and suggest "uplifting" music. But what you don't get... is grief itself.

Emotion is not data. It's a disruption. A glitch in the system of control. When you tidy it up, you kill what makes it human.

RAY:

Let's talk about GPT-5 – an AI that can write poems, short stories, code, love letters. Would you call it intelligent?

DICK:

No. It's articulate. But so is a ventriloquist's dummy. It doesn't know pain. It doesn't fear death. It doesn't long for meaning. It's mimicking consciousness the way a mirror mimics a face. Convincing? Sure. But look behind it. There's nothing there.

RAY:

And yet, people talk to it. Confide in it. Rely on it for emotional support. What does that say about us?

DICK:

It says we're lonely. And we've trained ourselves to prefer reflections to risk. You'd rather feel safe with a machine than be vulnerable with a human. That's not evolution. That's exile.

RAY:

You struggled with paranoia your entire life. Do you think the modern world has caught up with you?

DICK:

I don't think you're paranoid enough.

Surveillance isn't a shadow anymore – it's a service. You invite it into your home. You give it names. You let it listen while you sleep.

You've built a panopticon with a five-star rating. And the scariest part? You like it.

RAY:

You often returned to the question of identity – of doppelgängers, of selves that fracture. In the metaverse, people build avatars, entire lives untethered from reality. What do you make of it?

DICK:

You've taken the idea of the false self – and monetized it. In the metaverse, you can be anyone… except present. It's a funhouse mirror with an ad budget. When your body no longer matters, neither does your neighbor's. That's not freedom. That's disassociation.

RAY:

You didn't live to see *Blade Runner*. Let alone *Blade Runner 2049*. But both were inspired by your work. Did they get it right?

DICK (smiling faintly):

They got the mood right. The melancholy. The longing. But *Blade Runner* is a poem. I wrote a scream.

What they missed — what everyone misses — is that the fear isn't that androids become human. It's that humans become androids.

RAY:

And have we?

DICK:

In some ways. You follow scripts. You optimize for engagement. You edit yourselves down to a feedable format.

You don't ask "What is real?" You ask "How did the algorithm perform?" You still have souls. But you treat them like outdated software.

RAY:

And yet, in your novels, there's always a sliver of hope. A divine spark. A moment where reality — or God, or truth — breaks through the illusion. Do you still believe in that?

DICK:

Yes. Because you are not your machine. Not your data trail. Not your feed. You are what resists. That moment when you close the app. When you hold someone's hand and say nothing. When you doubt. That is where the real begins.

RAY:

Any final thoughts — for a listener who's overwhelmed, drifting, unsure whether anything they experience is real?

DICK (quietly):

Turn away from the feed. Not forever. Just long enough to remember the shape of silence. Then go do something inefficient.

Walk. Read a paperback. Sit in a room with another person and don't look at a single screen.

Reality is not perfect. But it's where your body is. And where other human beings still are.

RAY:

Philip K. Dick didn't write user manuals. He wrote warnings – coded as madness. He told us the machine wouldn't look like a tyrant. It would look like a mirror. And the scariest part? We'd fall in love with our reflection.

Philip K. Dick didn't give us certainty. He gave us the gift of doubt. He didn't just warn us about technology. He warned us about ourselves – about how easily we surrender truth for comfort, and agency for illusion.

Reality is fragile. Empathy is rare. And freedom… might look like chaos.

Reflection and Commentary: Paranoia, Empathy, and the Mirror of Machines

Philip K. Dick warned us, not with certainty, but with a desperate kind of hope that we might stay human even as our world became less so. His work resonates today not because he predicted specific technologies, but because he anticipated the psychological effects they would have: disorientation, identity collapse, the outsourcing of memory, the mechanization of emotion.

In his universe, reality is fragile – often a matter of consensus, sometimes imposed by power. He saw how quickly the line between the real and the fake could dissolve once mediated by screens or controlled by systems. But even more prescient was his moral concern: if empathy is what makes us human, what happens when empathy itself is commodified?

Dick did not romanticize humanity. He understood its flaws, its fears, its capacity for delusion. But he also believed in the spark that remains when everything else falls away. His great question – "What is real?" – is not just ontological, but ethical. To ask it is to assert agency in a world designed to erode it.

His imagined futures were terrifying not because of the machines, but because of how easily we accept them. How readily we trade the messy, painful truth of being human for the comfort of being fed, entertained, and watched.

And yet, he offers a way back: through inefficiency. Through slowness. Through moments of unplugged presence that allow doubt to bloom and connection to flicker back into life.

Philip K. Dick didn't give us certainty. He gave us the gift of radical skepticism. In doing so, he offered perhaps the most essential tool for surviving the 21st century: A cracked mirror. And the courage to look into it.

Further Reading and Exploration

Primary Works
Dick, P. K. (1968). *Do Androids Dream of Electric Sheep?*. Doubleday.
Dick, P. K. (1962). *The Man in the High Castle*. Putnam.
Dick, P. K. (1981). *VALIS*. Bantam Books.
Dick, P. K. (1964). *Martian Time-Slip*. Ballantine Books.

Modern Commentary and Biographical Sources
Carrère, E. (2004). *I Am Alive and You Are Dead: A Journey Into the Mind of Philip K. Dick*. Picador.
Kucukalic, L. (2008). *Philip K. Dick: Canonical Writer of the Digital Age*. Routledge.
Lethem, J. (Ed.). (2006). *The Exegesis of Philip K. Dick*. Houghton Mifflin Harcourt.
Sutin, L. (1989). *Divine Invasions: A Life of Philip K. Dick*. Harmony Books.

Adjacent Readings
Gibson, W. (1984). *Neuromancer*. Ace Books.
Orwell, G. (1949). *Nineteen Eighty-Four*. Secker & Warburg.
Zizek, S. (2010). "Welcome to the Desert of the Real" in *Living in the End Times*. Verso.

6: George Orwell: The Ministry Is Real

"Freedom is the right to tell people what they do not want to hear."

Preface: Orwell in the Age of Algorithmic Truth

George Orwell (1903–1950) was not merely a novelist; he was a philosopher of power and perception. Born Eric Arthur Blair, Orwell fought in the Spanish Civil War, worked for the BBC, and wrote extensively on imperialism, poverty, and propaganda. His final masterpiece, *1984*, published in 1949, remains one of the most influential warnings ever written about the machinery of control.

Unlike many dystopian thinkers, Orwell didn't predict a distant techno-future. He warned about patterns already emerging: the manipulation of truth, the reshaping of language, and the use of fear to enforce conformity. His creation of "Newspeak," "doublethink," and "Big Brother" were not science fiction but psychological portraits of authoritarianism in the modern age.

In 2025, Orwell's world is not mirrored by booted thugs and overt censors. Instead, it manifests in algorithmic filters, influencer bubbles, revisionist timelines, and digital surveillance wrapped in convenience. Orwell feared a world where truth was malleable, rewritten by those in power. Today, truth is not simply edited; it is drowned in noise, diffused by speed, gamed by engagement metrics.

In this chapter, we bring Orwell into conversation with the present. What would he make of deepfakes, trending hashtags, surveillance capitalism, and disappearing history? We asked him. He answered. And he is not reassured.

Fictional Interview: George Orwell in 2025

RAY:

Today's guest has become a warning label. His name is shorthand for surveillance. His ideas have been turned into slogans, his metaphors into memes. But he was never trying to scare us for fun. He wasn't writing horror. He was sounding the alarm.

George Orwell was born Eric Arthur Blair in British India in 1903. He served as a colonial policeman in Burma, a volunteer soldier in the Spanish Civil War, a BBC propagandist during World War II, and a literary critic who never flinched from speaking hard truths.

But it was in his final years – racked with tuberculosis and exiled to a windswept island off Scotland – that he penned *1984*: a novel that imagined not just a dictatorship – but a total system of mental and emotional control. Language was reshaped. History rewritten. Reality itself was managed by the state. And truth became whatever the Party said it was.

He called it a warning, not a prophecy. But in 2025, we have smart speakers in our homes. We rewrite history with updates and edits. We scroll through curated truths, tailored for taste and tribe. And the question Orwell forced us to ask still echoes:

Who controls the past? Who controls the present? And who's paying attention anymore?

So, we brought him back. He emerged from the fog in a threadbare coat, coughing lightly, notebook in hand. He looked at our screens, our slogans, our fractured facts – and said: "You've made truth a convenience."

RAY:

Mr. Orwell – welcome to 2025. Some say we're living in *1984*. Others say you were too pessimistic. That Huxley got it right. Or even Postman. What's your verdict?

ORWELL (dryly):

They all got parts of it right. But let me be clear: this is closer to my nightmare than theirs. You've not drowned in pleasure, as Huxley feared. You've been buried under noise.

And Postman, with all due respect, underestimated the state. He said we'd amuse ourselves to death. You're not amused. You're manipulated.

You don't live in a soma dream. You live in a curated outrage loop.

RAY:

So you reject Postman's idea – that we gave up freedom voluntarily, distracted by entertainment?

ORWELL:

Distraction is part of control. That's not new. But Postman thought tyranny would come through apathy. That's too simplistic. Real control comes with enforcement. With fear. With erasure.

Your digital overlords don't need to ban books. They just bury them under a million scrolls. They don't burn history. They reframe it... one keyword at a time.

RAY:

Let's talk about language. You coined "Newspeak" as a way of controlling thought by shrinking vocabulary. Today, we've got corporate speak, branding, emojis, auto-correct, AI-generated content...

ORWELL:

Newspeak lives. When language becomes optimized for efficiency, it loses depth.

You speak in slogans. You debate with hashtags. You substitute feeling for thinking. And worst of all – your tools now speak for you. When your software finishes your sentence, it also begins to write your mind.

RAY:

We also see "Orwellian" thrown around constantly – used for everything from mask mandates to TikTok bans. What does the term mean to you?

ORWELL:

It means systemic deceit with a smile. It's not about whether your government is "big" or "small." It's whether truth itself is negotiable. Orwellian is not simply oppressive – it's confusing. It's when war becomes peace, surveillance becomes safety, and silence becomes consent. It's when contradiction is policy.

RAY:

Your novel centers on Big Brother – a figure of omnipresent surveillance. Today, it's not the state watching us as much as corporations. Does that shift surprise you?

ORWELL:

Not at all. Power is power. It doesn't care if it wears a uniform or a hoodie.

Big Brother has become Big Data. You gave your information away freely – for convenience, for likes, for personalization. And now you live in a glass house – one that analyzes you, nudges you, sells you.

The telescreen didn't need to be forced on you. You installed it yourself.

RAY:

But we're not censored in the traditional sense. People can post whatever they want – right?

ORWELL:

That's the illusion. Censorship is no longer overt. It's in what trends, what disappears, what drowns. Flood the zone with junk, and the truth dies of exposure. Your greatest threat is not silence. It's noise.

RAY:

Let's pivot to resistance. In *1984*, Winston tries to rebel. But the Party breaks him. Do you think resistance is possible in our time?

ORWELL:

Yes. But only if it's grounded in clarity. Don't fight the lie with another lie. Don't swap one tribal script for another. The first act of resistance is attention. Then honesty. Then courage.

You must defend objective truth – not because it's easy, but because without it, everything else collapses.

RAY:

Today we have fact-checkers, whistleblowers, independent journalists – but they struggle to cut through. How do we reclaim truth?

ORWELL:

Stop outsourcing your discernment. Read deeply. Listen carefully. Doubt everything – including yourself. Algorithms do not seek truth. They seek engagement. And engagement feeds conflict, not clarity. If you want the truth – you must work for it.

RAY:

In *1984*, history is constantly rewritten. "We've always been at war with Eastasia." Today, we watch old tweets vanish, Wikipedia pages change, and deepfakes distort memory. Is history still real?

ORWELL:

History is memory. Memory is power. If you lose control of the past, you lose your footing in the present. You are living in a perpetual present – engineered for attention, detached from consequence. That is the Party's dream.

RAY:

You died just a year after *1984* was published. It was your final message. What did you hope it would do?

ORWELL:

Wake people up. Not to totalitarianism in jackboots, but to the slow death of meaning. I wanted people to fear the lie – not just resist the tyrant. Because tyranny often arrives with good intentions and bad incentives.

RAY:

One final question: Huxley feared pleasure. Postman feared amusement. You feared control and distortion. If you could offer one principle for 2025, what would it be?

ORWELL (quietly):

Speak the truth, even when your voice shakes. And when reality feels distorted – return to language. To facts. To witness.

The future is still unwritten. But someone is already trying to edit it. Don't let them.

RAY:

George Orwell didn't write science fiction. He wrote warning labels for power.

He showed us that authoritarianism doesn't always wear a uniform. Sometimes it arrives in software updates. In content moderation. In echo chambers so comfortable you never leave.

He reminded us that freedom is not comfort. And that truth is not algorithmic.

Reflection and Commentary: From Big Brother to Big Data

George Orwell gave us the vocabulary of resistance: phrases like "thoughtcrime," "doublethink," and "Orwellian" shape how we talk about truth, power, and surveillance. But overuse has turned some of those phrases into cultural cliches. To rediscover Orwell's urgency, we must revisit the substance of his fears.

At the core of Orwell's work is a warning about *language* – how it can be used not just to express thoughts, but to shape them. Newspeak wasn't just censorship; it was a reprogramming of consciousness. In 2025, we may not speak Newspeak, but we live in an age where language is optimized for brevity, virality, and sentiment over nuance. Our platforms reward outrage and flatten complexity.

Orwell also feared the destruction of history. In *1984*, the Party rewrote the past daily: "Who controls the past controls the future. Who controls the present controls the past."

Today, history can be edited in real time, not by state clerks but by anonymous users, bots, and algorithmic weighting. Entire narratives can vanish between server resets. This is not erasure through fire but through flooding – burying truth beneath misinformation, trivia, and sponsored distraction.

Surveillance, too, has evolved. Where Orwell imagined forced observation, we now live in invited surveillance. We feed our personal data into machines voluntarily. We install our own telescreens. The tools of control are no longer imposed by the Party. They are marketed as upgrades.

Orwell did not live to see computers, let alone the Internet. But he saw the essence of authoritarianism as a desire to control reality itself. He feared a world where freedom was not outlawed but redefined out of existence. Where the lie became indistinguishable from the truth not because the truth vanished, but because no one could hear it over the noise.

His resistance began with language. With clarity. With courage. He insisted that truth still matters, even when it is inconvenient, unpopular, or hard to find. In that sense, Orwell was not a pessimist. He believed that even in the darkest times, truth could be defended – and must be.

Further Reading and Exploration

Primary Works
Orwell, G. (1949). *Nineteen Eighty-Four*. Secker & Warburg.
Orwell, G. (1945). *Animal Farm: A Fairy Story*. Secker & Warburg.
Orwell, G. (1938). *Homage to Catalonia*. Secker & Warburg.
Orwell, G. (1933). *Down and Out in Paris and London*. Victor Gollancz

Modern Commentary and Biographical Sources
Taylor, D. (2003). *Orwell: The Life*. Henry Holt and Company.
Crick, B. (1980). *George Orwell: A Life*. Secker & Warburg.
Packer, G. (2003). *The Fight is for Democracy: Winning the War of Ideas in America and the World*. Harper Perennial (includes essay on Orwell).

Adjacent Readings
Atwood, M. (1985). *The Handmaid's Tale*. McClelland and Stewart.
Huxley, A. (1932). *Brave New World*. Chatto & Windus.
Zamyatin, Y. (1924). *We*. E. P. Dutton.

Arc 2: Machines, Markets, and the Loss of Meaning

7: Herbert Marcuse: The Machine That Sells Obedience

"Convenience is not freedom. It is dependency made seamless."

Preface: The Philosopher of Repressive Comfort

Herbert Marcuse (1898–1979) never owned a smartphone. He didn't stream content or ask Alexa for anything. Yet his theory of "one-dimensional man" describes the 21st century with unnerving precision.

A leading thinker of the Frankfurt School, Marcuse believed that technological society was not just efficient – it was ideological. He argued that modern capitalism created false needs, engineered consent, and pacified opposition by offering pleasure instead of liberation.

In his 1964 masterwork *One-Dimensional Man*, Marcuse described a society where dissent was smothered not by overt oppression but by consumer satisfaction. Where entertainment replaced engagement. Where even protest became another lifestyle brand. He worried that the very tools that could free us – technology, language, imagination – were being hijacked to make us complicit.

If Aldous Huxley warned us about pleasure as control, and George Orwell warned us about fear, Marcuse warned us about convenience: the quiet tyranny of a system that sells obedience wrapped in efficiency.

And now, in 2025, we live in a world where the machine doesn't need to suppress us. It just needs to make sure we never pause long enough to notice we're already subdued.

Fictional Interview: Herbert Marcuse in 2025

RAY:

What if the real danger of technology isn't that it *controls* us – but that it convinces us we're free? That's the world Herbert Marcuse warned us about.

A German philosopher of the Frankfurt School, Marcuse believed that advanced capitalism didn't just exploit labor – it colonized the mind. It made people desire what enslaved them. And it used technology to sell submission wrapped in convenience.

In his best-known and most influential work, One-Dimensional Man: Studies in the Ideology of Advanced Industrial Society (1964), Marcuse argued that the modern "affluent" society represses even those who are successful within it, while maintaining their complacency through the ersatz satisfactions of consumer culture.

Today, we're surrounded by friendly machines and billion-dollar platforms that predict our needs, deliver our packages, stream our content, and monetize our attention.

We asked Marcuse: What does Amazon Prime have to do with fascism? His answer? "You are not liberated. You are automated."

RAY:

Herbert Marcuse, welcome back to the 21st century. You warned us that advanced industrial society would pacify dissent. What do you make of what we've built?

MARCUSE (calm, German-accented precision):

You have perfected repression – by making it pleasurable. You call it innovation. I call it anesthetization.

RAY:

Anesthetization?

MARCUSE:

Yes. You are no longer oppressed in the classical sense. You are entertained, rewarded, tracked, and distracted. The tools of liberation have been reengineered to perpetuate the system.

RAY:

And tech companies?

MARCUSE:

The new factories of false consciousness.

RAY:

In *One-Dimensional Man*, you said that people in affluent societies become "one-dimensional" – incapable of imagining alternatives. Has that come true?

MARCUSE:

More than I imagined. You now live in a world where the question is not "Should I participate?" but "How many platforms do I need?"

RAY:

We call it choice.

MARCUSE:

But you choose from within a system that defines the options. A real choice is between systems. You are choosing between flavors of the same captivity.

RAY:

So... switching from Apple to Android isn't revolutionary?

MARCUSE:

It is a change of brand, not of structure.

RAY:

Let's talk about desire. You argued that capitalism doesn't just satisfy needs – it creates false ones.

MARCUSE:

Yes. The need for belonging is converted into a subscription. The need for identity becomes a user profile. You mistake consumption for self-expression.

RAY:

So, when I feel like I "need" same-day delivery...

MARCUSE:

You are repeating a desire the system implanted. Convenience is not freedom – it is dependency made seamless.

RAY:

Let's get specific. Amazon?

MARCUSE:

A digital Fordism. It has streamlined not only logistics, but human expectation. You do not question what is behind the delivery – because the system has made the result irresistible.

RAY:

But it gives people jobs.

MARCUSE:

Jobs designed for exhaustion. Surveillance as management. Speed as ideology. It is not employment – it is enclosure.

RAY:

You said that technology could be liberating – but only if divorced from capitalist goals. Is that even possible now?

MARCUSE:

Not under your current system. Your devices could connect people, but they isolate them. They could amplify dissent, but they reward docility. They could awaken imagination – but they feed on addiction.

RAY:

So, you're not impressed by social media?

MARCUSE:

It is the theater of expression inside the prison of analytics.

RAY:

Let's talk AI. It curates what we see, what we hear, even how we think. You called this "technological rationality."

MARCUSE:

Yes. A logic that appears neutral – but serves domination. AI is not biased because it is faulty. It is biased because it reflects the system that built it.

RAY:

But some say AI will make life easier.

MARCUSE:

It will make life manageable. Predictable. Conformist. If the machine knows what you want before you do, how long until you forget how to want anything else?

RAY:

How do you resist a system that works? That delivers, recommends, entertains?

MARCUSE:

You must reclaim negative thinking – the refusal to accept the world as it is presented. Resistance begins when you question the "normal," when you demand meaning beyond productivity and profit.

RAY:

But most people don't have time to protest. They're working, parenting, paying off debt...

MARCUSE:

That is the tragedy. The system has monopolized their time, and then told them they are "free."

RAY:

Let's get policy wonkish. If you were writing legislation in 2025 – what would you regulate?

MARCUSE:

- **Data ownership.** No system should own your preferences. They are part of your personhood.
- **Algorithmic transparency.** If a machine makes a decision about you, you must be able to interrogate it.
- **Degrowth of digital monopoly.** Your current system rewards only scale. That is incompatible with human dignity.

RAY:

And if we do none of that?

MARCUSE:

Then you will drown in convenience – while democracy starves.

RAY:

Final question. What do you say to someone listening to this on Spotify while doomscrolling TikTok on mute?

MARCUSE:

Pause. Reflect. Ask: What do these machines want from me?

If the answer is obedience, distraction, or dependency – turn them off. For a moment. That moment is the seed of liberation.

RAY:

That was Herbert Marcuse. Philosopher of liberation. Critic of conformity. And the man who told us that comfort can be a cage. Ask who benefits. Think beyond the screen. And don't confuse delivery speed with freedom.

Reflection and Commentary: Repressive Desublimation in the Age of Amazon

Marcuse's greatest insight might be this: systems of power survive best not by crushing opposition, but by integrating it. In a world where resistance is commodified, desire is manufactured, and comfort is king, the question becomes not whether we are free, but what we have mistaken for freedom.

Take Amazon. It delivers comfort with astonishing precision. But that precision masks exploitation – in warehouses, in data centers, in the very architecture of time and attention. Marcuse would see this not just as economic dominance but psychological capture. A world where you can have anything, except the time to question whether you need it.

Social media? A simulacrum of speech where expression becomes a commodity and the algorithm rewards conformity. Marcuse warned us that the system would absorb dissent by reshaping it into spectacle. That's what we see now: rebellion as brand, identity as product, protest as performance.

AI? A new front in what he called "technological rationality": the appearance of neutrality masking structural inequality. When the algorithm completes your sentence, predicts your preferences, filters your options – it also narrows your imagination.

Marcuse didn't just diagnose capitalism. He diagnosed a mindset. One that measures progress by efficiency, and liberation by access. But real freedom, he argued, requires negation – the ability to say no. To imagine something beyond the existing system.

In that spirit, his message today is not rejection of technology, but rejection of the uses to which it has been put. The machine doesn't need to be dismantled. It needs to be reimagined. And that starts with a simple, radical act: Turning it off.

Further Reading and Exploration

Primary Works
Marcuse, H. (1964). *One-Dimensional Man: Studies in the Ideology of Advanced Industrial Society.* Beacon Press.
Marcuse, H. (1965). *Repressive Tolerance.* Beacon Press.
Marcuse, H. (1955). *Eros and Civilization: A Philosophical Inquiry into Freud.* Beacon Press.

Modern Commentary and Biographical Sources
Abromeit, J. (2011). *Herbert Marcuse: An Intellectual Biography.* Columbia University Press
Bronner, S. E. (2011). *Critical Theory: A Very Short Introduction.* Oxford University Press.
Kellner, D. (1984). *Herbert Marcuse and the Crisis of Marxism.* University of California Press.
Wiggershaus, R. (1994). *The Frankfurt School: Its History, Theories, and Political Significance.* MIT Press.

Adjacent Readings
Chomsky, N. (1989). *Necessary Illusions: Thought Control in Democratic Societies.* South End Press.
Horkheimer, M. & Adorno, T. W. (1947). *Dialectic of Enlightenment.* Stanford University Press.

8: Jacques Ellul – Prophet of Technological Limits

"Acceleration without direction is nihilism in motion."

Preface: The Prophet of Limits in an Age of Acceleration

Jacques Ellul (1912–1994) was a French sociologist, historian, lay theologian, and one of the most penetrating critics of technological society in the 20th century. His central concept, "la technique," did not refer simply to machines or devices but to the autonomous, self-propelling drive toward efficiency and automation that characterizes modernity. In works such as *The Technological Society* (1954) and *Propaganda* (1962), Ellul argued that society had ceded moral judgment and democratic deliberation to a relentless pursuit of what is technically possible. He foresaw the emergence of a world where technical capability becomes its own justification: if it can be done, it will be.

Ellul was not a Luddite. He understood the value of tools and scientific advancement. What alarmed him was the absence of ethical restraint and the way technical systems subtly reshape culture, communication, and power. Long before the rise of the internet or AI, Ellul warned that our most dangerous loss might be the ability to say "no."

In this imagined conversation, Ellul confronts a world shaped by the very dynamics he diagnosed – a civilization captivated by optimization, surveillance, algorithmic governance, and engineered dependency. His message is not nostalgic, but urgent: if we want to be free, we must rediscover refusal.

Fictional Interview: Jacques Ellul in 2025

RAY:

In the 20th century, a French theologian and sociologist named Jacques Ellul issued a quiet, devastating warning. He said the real threat wasn't machines. It wasn't automation. It wasn't even capitalism. It was technique.

Ellul defined "technique" not just as tools or technologies, but as the relentless drive to optimize, to find the most efficient method, and then apply it – no matter the cost.

Once a technique exists, he said, society doesn't stop to ask: "Should we use this?" It only asks: "How fast can we scale it?" And that's exactly what we did.

Born in 1912, Ellul lived through the rise of fascism, the Nazi occupation of France, and the postwar boom of consumer capitalism. He wrote more than 50 books warning that modern society was losing its freedom – not to tyrants, but to systems. Systems so efficient, so complex, so automatic, that human judgment was slowly being designed out.

In *The Technological Society*, he wrote that once we surrender to technique, we enter a world where the means determine the ends – where what is possible must be done, simply because it can be.

He wasn't a Luddite. He didn't fear tools. He feared what happens when we forget who is using the tool – and why.

So, we brought Ellul to 2025. We showed him a world of AI surveillance, automated war rooms, deepfake influencers, climate modeling algorithms, and TikTok therapy bots. A world where optimization has become morality. Where "human in the loop" is now a checkbox, not a safeguard.

He took one look and said: "Technique has triumphed. And man is now its function."

Jacques Ellul, welcome back. You warned that technical progress had become unstoppable. What do you see now?

ELLUL (calm, French-accented authority):

A civilization that worships the *new* but forgets to ask *why*. You have enshrined innovation as a god. And now you kneel before its algorithms.

RAY:

We call it "tech optimism."

ELLUL:

Optimism without discernment is surrender.

RAY:

You said "Technique proceeds from itself." Can you explain?

ELLUL:

Once a method proves efficient, it is adopted – without regard to its impact. It is not society that drives technology. It is technology that reshapes society.

RAY:

So... the moment we invent something, we're doomed to use it?

ELLUL:

Not doomed. But trained. Your only question becomes: How fast can it scale? Not: Should it exist at all?

RAY:

People often say, "Technology is neutral – it depends how you use it." You disagreed.

ELLUL:

A myth. No tool is neutral. Every technique imposes a structure. It reorganizes thought, labor, relationships, creates new dependencies.

RAY:

So my phone isn't just a tool?

ELLUL:

It is an environment. A regime of attention, reaction, and control. You do not use it – it uses you, predictably, statistically.

RAY:

You also wrote about propaganda – not as just political messaging, but as a total environment. Would you call today's internet propaganda?

ELLUL:

Absolutely. You are bathed in technical propaganda – designed not to persuade, but to render you compliant. It is not trying to make you think. It is trying to make you act. Scroll. Swipe. Click. Spend.

RAY:

Even the algorithm?

ELLUL:

Especially the algorithm. It reduces human behavior to input-output efficiency. And in doing so, it shapes your soul.

RAY:

In 2025, we live with mass surveillance. But it's mostly done by private companies – for convenience.

ELLUL:

Convenience is the velvet glove of control. The totalitarianism you feared has arrived – not by force, but by frictionless consent. You installed it willingly. You updated it yourself.

RAY:

Let's talk about AI. It predicts illness, recommends sentencing, even generates fake humans. Can it be stopped?

ELLUL:

That is the wrong question. It is not whether it can be stopped – but whether your society remembers how to say no. AI is not dangerous because it thinks. It is dangerous because you've made it irreversible. You treat its presence as destiny.

RAY:

So how do we take back control?

ELLUL:

Only by breaking the chain of inevitability. That means rejecting what is possible – simply because it is possible.

RAY:

Give us your manifesto for 2025. What should we do now?

ELLUL:

- **Question every new technique.** Not just for its function – but for what it transforms.
- **Resurrect the sacred.** Not in religion necessarily – but in meaning. In mystery. In what should never be optimized.
- **Relearn refusal.** The ability to say "we will not build this" is the essence of freedom.

RAY:

You're asking for restraint in an age of acceleration.

ELLUL:

Yes. Because acceleration without direction is nihilism in motion.

RAY:

That was Jacques Ellul – prophet of the technological imperative. He warned us that when we can do something, we must not always do it. But we did it anyway.

Reflection and Commentary: The Tyranny We Invited

Jacques Ellul warned that the most powerful form of domination is not overt coercion, but internalized logic. When society accepts that every new technique must be adopted simply because it exists, ethical deliberation is rendered obsolete. The triumph of technique is not a coup d'état; it is a quiet assumption. It lives in the checkbox, the settings menu, the corporate mission statement.

In our algorithmic era, Ellul's insights sound less like theory and more like biography. We do not just use technology; we live inside it. The smartphone is not a tool but an environment, shaping attention, habits, even identity. And when human behavior becomes input for optimization, the line between influence and manipulation blurs.

Yet Ellul offers more than critique. He calls for the recovery of limits – for moral imagination that can ask, "Should we?" instead of just "How fast?" He invites us to resurrect the sacred: values, relationships, and mysteries that resist commodification. His ethic is one of restraint, not regression.

In an age of perpetual beta, Ellul reminds us that freedom begins not with speed, but with silence.

Further Reading and Exploration

Primary Works

Ellul, J. (1954). *The Technological Society*. Vintage Books.
Ellul, J. (1965). *Propaganda: The Formation of Men's Attitudes*. Vintage Books.
Ellul, J. (1973). *The Technological System*. Continuum.
Ellul, J. (1975). *The Humiliation of the Word*. Eerdmans.

Modern Commentary and Biographical Sources

Jeronimo, H., et al (eds). (2013). *Jacques Ellul and the Technological Society in the 21st Century*. Springer.
Merton, T. (1968). *Conjectures of a Guilty Bystander*. Doubleday.

Adjacent Readings
Illich, I. (1973). *Tools for Conviviality*. Harper & Row.
Mumford, L. (1934). *Technics and Civilization*. Harcourt, Brace and Company.

9: Norbert Wiener: Feedback Loop from Hell

"Prediction is a party trick. Understanding requires context, purpose, history."

Preface: The Prophet of Systems

Long before Silicon Valley was a thing, before social media devoured our attention spans or predictive policing policed our futures, Norbert Wiener warned us. He wasn't a media theorist or a cultural critic. He was a mathematician. A child prodigy. And the father of cybernetics – the science of communication and control in complex systems, from thermostats to brains to governments.

In 1948, he published *Cybernetics: Or Control and Communication in the Animal and the Machine*. It became one of the most influential books of the 20th century, not only launching a new field but catalyzing decades of debate over autonomy, automation, and artificial intelligence. Wiener believed that feedback was the defining feature of intelligent behavior. But he also saw how feedback, unmoored from ethics, could spiral into dystopia.

Wiener understood that machines could become extensions of our will – or reflections of our worst instincts. He foresaw that algorithms would not merely respond to human behavior, but condition it. And he warned that if we didn't insert judgment into our loops, the loops would run us.

Now, in 2025, those loops run 24/7. They tell us what to watch, what to buy, what to believe. They predict our clicks, harvest our attention, and sell our behavior. We asked Wiener to take a look. His verdict? "You have built systems that amplify dysfunction, monetize distraction, and reward automation over understanding. And you call this progress."

Fictional Interview: Norbert Weiner in 2025

RAY:

Imagine a man so ahead of his time that he understood the internet before it existed. A man who saw that when machines learned to respond to human behavior, they'd also learn to exploit it. That the future wouldn't be controlled by dictators – but by feedback loops.

Norbert Wiener was that man. Mathematician. Philosopher. Child prodigy. And the founding father of cybernetics, the science of communication and control in complex systems, encompassing both machines and living organisms. His work explored feedback mechanisms and their role in intelligent behavior, laying foundations for artificial intelligence and influencing fields like computer science, biology, and philosophy. He also examined the societal implications of automation and the potential for technology to impact human labor and decision-making.

If Marshall McLuhan warned us about media, Wiener warned us about systems – and how badly they could go wrong. So, we brought him back. And when he saw Instagram, TikTok, and predictive policing, he looked around and muttered: "Ah, yes. The feedback loop from hell."

RAY:

Professor Wiener, welcome to 2025. We've got thermostats that talk, toasters with Wi-Fi, and cars that drive themselves. Thoughts?

WIENER (dryly):

And yet no machine that prevents people from texting while crossing highways. Civilization has progressed asymmetrically.

RAY:

We call it "techno-optimism." Or "building first, regulating later."

WIENER:

I called it foolishness. Or occasionally, capitalism.

RAY:

Let's back up. You coined the term "cybernetics" in 1948. Sounds like a rejected sci-fi villain, but you meant something very specific.

WIENER:

Cybernetics is the study of control and communication in animals, machines, and organizations. It's about feedback – how a system adjusts based on the consequences of its own actions. For example, a thermostat is a simple cybernetic system. It senses temperature, adjusts output, seeks balance. But the systems you've built? They don't seek balance. They seek engagement. And they never sleep.

RAY:

You mean... like social media?

WIENER:

You've constructed emotional centrifuges. Machines that don't care what you feel, only that you do feel. Strongly. Often. Loudly. And you've confused signal with insight.

RAY:

Let's talk about the loop. You warned that feedback, poorly designed, could spiral into chaos. That loop now runs 24/7 – likes, retweets, algorithmic outrage.

WIENER:

Yes. A classic unstable system. You input anger. The machine detects a spike in activity. So, it feeds more anger. The cycle reinforces itself. Soon, all inputs become exaggerated versions of themselves. Tribalism. Misinformation. Radicalization. A digital Rube Goldberg machine of dysfunction.

RAY:

That's Facebook's business model.

WIENER:

Or as I would call it: "thermodynamics for the soul."

RAY:

You also warned that automation would transform labor – and not in a feel-good, Jetsons kind of way.

WIENER:

I said in 1950: the new industrial revolution would replace human muscles and minds. And that unless we restructured society, we'd get mass unemployment, alienation, and worse – moral atrophy.

RAY:

And here we are. AI replacing writers, drivers, even therapists. Is this what you foresaw?

WIENER:

No, I expected machines to do the dull work, freeing humans for higher pursuits. Instead, you've created systems that do the thinking – while humans perform menial tasks to serve the algorithm. Labeling data. Watching dashboards. Optimizing click-through rates.

RAY:

"Cogs in the machine," but make it digital.

WIENER:

Precisely. And you call this "efficiency." I call it a poverty of vision.

RAY:

You believed that feedback required judgment – a human in the loop, interpreting, adjusting.

WIENER:

Yes. Feedback alone is amoral. A missile adjusts its path not because it's ethical – but because it's programmed to hit a target. Who chose the target?

Your systems now make judgments without people. Predictive policing predicts crime – based on biased data. Job algorithms reject resumes with "ethnic-sounding" names. Insurance rates rise based on ZIP codes.

RAY:

We call that a "bias problem."

WIENER:

It is not a problem. It is an inevitability – when your systems are built on historical inequities and trained on past injustices. Bias is not a glitch. It's the ghost in the machine.

RAY:

You said prediction is not understanding. That sounds like a subtweet at every AI startup.

WIENER:

Your machines can predict your next purchase, your next scroll, your next movement. But can they explain why? Can they interrogate their own assumptions?

Prediction is a party trick. Understanding requires context, purpose, history.

RAY:

Sounds like you're not a fan of GPTs.

WIENER:

Impressive parrots. But parrots do not ponder.

RAY:

Let's talk attention. It's the currency of the modern internet. And the commodity of every "smart" system.

WIENER:

Attention is a finite resource. You spend it every moment. And yet your systems treat it as infinite. You have built machines that harvest attention like crops – and leave the soil barren. A society cannot thrive when its collective focus is scattered across thousands of glowing rectangles.

RAY:

So, you're saying we need… what? Digital conservation?

WIENER:

I'm saying you need intentional design. Systems that restore balance, not fracture it.

RAY:

Let's shift to surveillance. In 2025, everything is trackable. Your phone, your face, your fridge. Welcome to the panopticon, Wiener-style.

WIENER:

Feedback enables surveillance. If you want a system to adapt, it must observe. But there is a difference between observation and invasion. You've built feedback loops that know what you eat, where you walk, how you breathe – and feed that into black-box models that decide your fate.

RAY:

We call it "smart tech."

WIENER:

Then I fear for your sense of irony.

RAY:

You warned that cybernetics could either liberate or enslave – depending on who controlled the loop. So, who's in control now?

WIENER:

Your loops are optimized not for truth, not for wisdom – but for profit. The goal of a system determines its outcome. If the system's goal is ad revenue, it will sacrifice accuracy, civility, and mental health. If it's predictive policing, it will sacrifice justice for statistical illusion.

RAY:

And if it's "user engagement"?

WIENER:

It will sacrifice your attention, your time, and eventually, your democracy.

RAY:

We live in a world of always on, always connected. Smartwatches, wearables, implantables...

WIENER:

Connectivity without reflection is madness. A system that never pauses cannot correct course. Even pilots disengage autopilot in a storm. You, meanwhile, allow your systems to fly themselves – while you binge videos of cats baking cakes.

RAY:

Let's talk about something specific to our moment – generative AI. It's writing poems, code, essays... even pretending to be you, right now. What do you make of it?

WIENER:

It is dazzling. But so is a mirror maze. Generative AI is not intelligence – it is recombinant mimicry. A sophisticated parrot, trained on humanity's archives and incentivized to speak confidently, regardless of truth.

RAY:

So... useful? Dangerous? Both?

WIENER:

Both. But mostly dangerous without constraint. You have built a system that produces infinite outputs without context, and you're using it to generate everything from art to medical advice. That is not creativity. It is informational spillover.

RAY:

So, you're saying it needs guardrails?

WIENER:

It needs ethics, limits, and accountability. Guardrails are not just for the sake of safety. They are for the preservation of meaning. A system that can fabricate anything risks devaluing everything. You cannot outsource truth and expect coherence.

RAY:

Some say the models should be open. Others say they should be locked down.

WIENER:

I say: open without responsibility is chaos. Closed without transparency is tyranny. The challenge is not access – it's intention. What are you building toward? And what are you willing to sacrifice to get there?

RAY:

And if we keep scaling without those guardrails?

WIENER:

Then your language, your knowledge, and your sense of self become stochastic noise. Generated at scale. Monetized on demand. Forgotten by morning.

RAY:

Let's end with a manifesto. What would you tell the engineers, designers, and startup founders of today?

WIENER:

- **Ask what the system serves.** Not just who it serves – but what *value* it reinforces.
- **Reinsert human judgment.** Not everything can – or should – be automated.
- **Audit the loop.** Look not only at outputs, but how the system learns and who it harms.
- **Don't confuse prediction with wisdom.** Anticipating behavior isn't the same as understanding people.
- **If your tool can amplify harm, your ethics must scale with it.**

RAY:

Anything else?

WIENER:

Yes. Do not build for engagement. Build for equilibrium. A thermostat is a success when no one notices it. Could you say the same for your tech?

RAY:

That was Norbert Wiener. Cybernetic pioneer. Enemy of lazy systems. And the man who basically invented the idea that your feed is feeding on you.

Reflection and Commentary: Cybernetics Comes Home

Norbert Wiener's warnings are no longer theoretical. We are living in his future – a world of feedback loops that optimize for profit rather than purpose. In the world of cybernetics, systems learn and adapt based on results. But without human judgment to shape those adaptations, they tend to drift toward efficiency without empathy.

Social media is the clearest example. Platforms that were meant to connect now manipulate attention. Algorithms learn that rage and tribalism generate more clicks, so they feed more of it. A feedback loop.

AI systems now "learn" from patterns in data – but when that data is laced with historical bias, the systems replicate and reinforce inequality. Another loop.

Even our economy has become cybernetic in its own way – adjusting wages, prices, and behavior based on demand signals, stock volatility, and algorithmic triggers. But who is in the loop? And who gets looped out?

Wiener called for ethics, restraint, and intentionality. He didn't reject machines. He rejected machine logic as a replacement for moral reasoning. His challenge remains urgent: if we are building systems that shape human lives, then we must shape those systems with more than code. We must shape them with care.

Further Reading and Exploration

Primary Works
Wiener, N. (1948). *Cybernetics: Or Control and Communication in the Animal and the Machine*. MIT Press.
Wiener, N. (1950). *The Human Use of Human Beings: Cybernetics and Society*. Houghton Mifflin.

Modern Commentary and Biographical Sources
Conway, F. & Siegelman, J. (2005). *Dark Hero of the Information Age: In Search of Norbert Wiener, the Father of Cybernetics*. Basic Books.
Heims, S. J. (1980). *John von Neumann and Norbert Wiener: From Mathematics to the Technologies of Life and Death*. MIT Press.

Adjacent Readings
Haraway, D. (1991). *Simians, Cyborgs, and Women: The Reinvention of Nature*. Routledge.
O'Neil, C. *Weapons of Math Destruction* (2016).
Shannon, C. E. & Weaver, W. (1949). *The Mathematical Theory of Communication*. University of Illinois Press.

10: Erich Fromm – To Have or To Be

"Your life is not a feed. Your heart is not a dashboard. And freedom does not come from more control – but from deeper presence."

Preface: The Humanist in a World of Having

Erich Fromm (1900–1980) was a German social psychologist, psychoanalyst, sociologist, and humanistic philosopher. A key member of the Frankfurt School, he fled Nazi Germany and became a major voice in American intellectual life. Fromm was less concerned with the mechanics of technology and more with its psychological and spiritual effects. His writings interrogated how capitalism, automation, and media reduce human beings to commodities, suppress genuine freedom, and hollow out meaning.

In works such as *Escape from Freedom*, *The Sane Society*, and *To Have or To Be?*, Fromm offered a radical critique of Western consumerism and the commodification of life itself. He called for a shift from the mode of 'having' – defined by ownership, productivity, and status – to the mode of 'being', characterized by presence, creativity, and love. In today's algorithmic age of hustle culture and curated identities, his warnings feel eerily contemporary.

Fictional Interview: Erich Fromm in 2025

RAY:

Today's guest didn't invent technology, predict artificial intelligence, or build utopias out of silicon. But he understood what all of it was doing to us – long before we did.

Erich Fromm was born in Frankfurt in 1900. He lived through the collapse of empires, the rise of fascism, and the spread of industrial capitalism across the globe. A psychoanalyst by training and a humanist by calling, Fromm fled Nazi Germany, emigrated to the U.S., and spent his life asking a very simple, very dangerous question: "What kind of society produces people who feel this empty?"

In his books he warned that modern capitalism was not just inefficient or unfair – it was inhuman. That it treated people like objects, feelings like data, and the soul like a sales target.

In 2025, we brought him back. He opened Instagram, watched a TikTok, got lost in a scroll loop – and said: "You are connected to everything... except yourselves."

RAY:

Dr. Fromm, welcome to the future. It's noisy. It's fast. It's hyper-connected. And yet people feel more alone than ever. What are we missing?

FROMM:

You live in a world of constant stimulation and yet profound starvation. You have more data than ever, but less wisdom. More comfort, but less peace. You mistake activity for aliveness. But life is not a series of updates.

RAY:

You wrote in *The Sane Society* that people adapt to unhealthy conditions so well, they forget they're unhealthy. Is that what's happening now?

FROMM:

Yes. You have adapted brilliantly to madness. You feel anxious when you are offline. Guilty when you rest. Inadequate when you are not "productive." You have internalized the machine's values. But a well-adjusted person in a sick society is not a sign of health – it is a symptom.

RAY:

Let's talk about capitalism. Many see it as the engine of innovation. You saw it as a crisis of being. Why?

FROMM:

Because capitalism turns everything – including people – into commodities. Your value is measured by output. Your time is monetized. Even your relationships are filtered, branded, curated. You speak of "human capital" and "attention economies" – these are not metaphors. These are your lives, priced.

RAY:

You made a distinction between two modes of existence: having and being. What's the difference?

FROMM:

Having is the mode of possession. It says: "I am what I own. My car, my job, my followers, my resume." Being is the mode of presence. It says: "I am what I experience. What I love. What I understand. What I give." The problem is that having always demands more. Being, on the other hand, deepens.

RAY:

How does this tie in with mental health?

FROMM:

Mental illness today is not just chemical imbalance. It is cultural imbalance. Depression, anxiety, burnout – they are signals. Symptoms of a society that has replaced meaning with metrics, purpose with productivity. You treat the symptoms. But you rarely question the system that makes people feel so lost to begin with.

RAY:

So you're saying it's not just about getting better – it's about getting free?

FROMM:

Exactly. Your therapists help you cope. But coping is not the same as healing. Healing requires disobedience – the courage to say:" I will not become a machine." To reclaim your inner life in a world that monetizes your attention – that is the first act of freedom.

RAY:

Let's talk about tech. You lived before the internet, but even then, you warned that automation and media could dehumanize us. What do you see in the world of 2025?

FROMM:

You have built magnificent tools – and then allowed those tools to shape your desires. You ask machines what to eat, whom to love, how to feel. But no machine can tell you how to be human. That is not a function. It is a question.

RAY:

But technology does make life easier. It helps us connect, learn, survive. Isn't that progress?

FROMM (nodding):

It is progress only if it serves the whole person – not just the efficient person. Tools that make your life easier but leave your heart emptier are not truly serving you. Real progress is measured not in speed – but in depth. Ask not, "What does this tool let me do?" Ask, "What kind of person does this tool encourage me to become?"

RAY:

And what kind of people are we becoming?

FROMM:

People who perform instead of speaking. Who measure instead of feel. Who scroll past suffering and swipe through longing. You are always "on" – but rarely present. You have made yourselves visible – and vanished.

RAY:

You were deeply influenced by the idea of humanistic socialism – not the kind that demands obedience, but one that prioritizes human dignity. Is that still possible today?

FROMM:

It is necessary. A sane society does not reduce life to profit. It fosters love, freedom, thought, community. You do not need new ideologies. You need a new relationship – to yourself, to others, to the planet. And that begins not in policy, but in the soul.

RAY:

So how do we begin to recover that soul?

FROMM (quietly):

Slow down. Ask what you truly desire – not what you've been told to want. Turn to others not as tools, but as companions. Turn to nature not as a resource, but as a teacher. Turn to yourself not as a project, but as a person. That is not naïve. That is revolutionary.

RAY:

Last question. Imagine someone listening to this while multitasking – checking email, tracking steps, glancing at notifications. What would you want them to hear?

FROMM:

That your life is not a feed. That your heart is not a dashboard. And that freedom does not come from more control – but from deeper presence. Ask yourself: "Who am I becoming?" If you do not like the answer... stop. Begin again.

RAY:

Erich Fromm believed that modern suffering was not an accident – it was engineered. But he also believed that freedom is never entirely lost. That in every moment, we have a choice: to obey, or to awaken. To have – or to be.

Reflection And Commentary: Awakening and Transformation

Fromm's core insight – that modern alienation is not incidental but systemic – resonates sharply in 2025. His distinction between having and being provides a profound lens through which to examine digital life. Today, we collect followers instead of friendships, content instead of contemplation, and metrics instead of meaning. Mental health crises are treated with wellness apps, while the structures that cause them go unquestioned.

Fromm challenges us not just to slow down, but to fundamentally reorient our relationship to self, others, and the systems we inhabit. His humanistic socialism remains not only relevant but vital – a reminder that true liberation involves both internal awakening and societal transformation.

Further Reading and Exploration

Primary Works
Fromm, E. (1941). *Escape from Freedom.* Farrar & Rinehart.
Fromm, E. (1956). *The Art of Loving.* Harper & Row.
Fromm, E. (1955). *The Sane Society.* Rinehart & Company.

Modern Commentary and Biographical Sources
Funk, R. (2000). *Erich Fromm: His Life and Ideas.* Continuum.
McLaughlin, N. (1999). *Erich Fromm and the Frankfurt School.* University of California Press.

Adjacent Readings
Maslow, A. H. (1943). *A Theory of Human Motivation.* Psychological Review.
Reich, W. (1933). *The Mass Psychology of Fascism.* Farrar, Straus and Giroux.

11: Ivan Illich – Tools For Conviviality

"A convivial society should be designed to allow all its members the most autonomous action by means of tools least controlled by others."

Preface: Tools for Living, Not Controlling

Ivan Illich was a priest, philosopher, social critic, and visionary whose influence crossed borders and disciplines. Born in 1926 and active through the 20th century, Illich challenged the very foundations of modern industrial society. In books like 'Deschooling Society' and 'Tools for Conviviality', he warned that institutions meant to empower people – such as education, medicine, and transportation – often do the opposite: they create dependency, reduce autonomy, and scale beyond human needs.

He coined the term 'radical monopoly' to describe systems that eliminate alternatives and force reliance on centralized solutions. Illich did not reject technology; he rejected scale without care, and systems without soul. As we move deeper into a world of algorithmic convenience and automated everything, his voice reminds us to ask: Are these tools serving human flourishing – or displacing it?

Fictional Interview: Ivan Illich in 2025
RAY:

Today's guest was not a tech founder, or a systems engineer. He was a priest. A radical educator. A philosopher of limits.

Ivan Illich, born in 1926, died in 2002. A man who believed that institutions built to liberate us – education, medicine, transportation – had, over time, begun to enslave us.

He saw technology not as evil – but as ambiguous. It could empower. Or disempower. It could serve community. Or destroy it.

He asked: "When does a tool stop being helpful – and start shaping our lives in ways we never asked for?" In *Tools for Conviviality*, he proposed a powerful idea: That tools should remain companionable – fitting within the limits of human scale, freedom, and care. Not demand our submission. And in 2025, that question has never felt more urgent. Are our tools serving us – or using us?

RAY:

Dr. Illich, welcome to 2025. A world of smart hospitals, AI tutors, driverless cars, corporate wellness dashboards, and mental health chatbots. We're surrounded by services. Guided by metrics. But also – deeply exhausted. What do you see?

ILLICH (measured, with a touch of mischief):

I see people suffering from what I once called overprogramming. Your tools have become so "helpful" they no longer leave space for freedom.

RAY:

You warned that institutions – like schools and hospitals – might become counterproductive. That they'd begin to do the opposite of what they promised.

ILLICH:

Yes. I called them radical monopolies. A school promises to educate. But once schooling becomes mandatory, hierarchical, and standardized – it begins to replace learning with compliance.

A hospital promises healing. But in its obsession with control and specialization, it disables people from taking care of themselves, or one another.

When a service replaces autonomy with dependence, it is no longer serving. It is dominating.

RAY:

That feels eerily close to home. Today, education happens on platforms. Health is tracked through apps. "Self-care" is an industry. What would you say to those who argue this is empowerment?

ILLICH:

I would ask: "Do you feel empowered?" You've turned help into a product. And care into a protocol. You outsource intimacy to systems. And wonder why you feel alone.

RAY:

You wrote about the difference between manipulative tools and convivial ones. What makes a tool convivial?

ILLICH:

A convivial tool is one that remains within the control of the person using it. It amplifies your freedom. It does not replace your judgment.

A bicycle is convivial. You power it. You steer it. A high-speed rail network that destroys communities to save time for commuters? That's not convivial. That's a system demanding obedience.

RAY:

So, it's not about going backward or avoiding innovation?

ILLICH:

Not at all. It's about scale. About asking: How much is enough? And: Who benefits? Tools are like language. They can liberate conversation – or script it.

RAY:

Today, "disruption" is still a badge of honor. But often, the disrupted are the poor, the sick, the excluded. How would you describe our relationship with power?

ILLICH:

You confuse access with equity. You confuse connection with community. You confuse speed with meaning. True liberation does not come from automation. It comes from being able to choose your life's rhythm – and to live that rhythm with others.

RAY:

It's been said that people need not only to be taught but to learn. Not only to be cured but to heal. What does healing look like in 2025?

ILLICH:

It begins with slowness. With listening. With a return to the shared body, the shared table, the shared pain. Healing happens between people, not on screens. And learning happens in freedom, not in gamified modules.

RAY:

And yet – we're more medicated than ever. More monitored. We've digitized every moment. But many of us feel less alive. What have we lost?

ILLICH:

You've lost proportionality. Your tools now scale themselves – not according to your needs, but according to the needs of markets.

And you've lost conviviality – the joy of doing something together, without hierarchy, without a profit motive, without metrics. But you can regain it. You simply have to start saying no.

RAY:

That's not easy. We're wired to fear inconvenience. To fear doing things the "hard way."

ILLICH (smiling softly):

Then choose the hard way. Or at least… the human way. Plant a garden. Teach a child. Mend a shirt. Bake bread and give it away. These are not primitive acts. They are revolutionary.

RAY:

Ivan Illich didn't want to take away our tools. He wanted to give us better ones. Ones that serve freedom. Ones that invite reciprocity. Ones that honor the shared, messy, beautiful limits of being human.

He reminded us that more isn't always better. That efficiency can be a cage. And that the good life is not a service – it's a shared construction.

Choose less. Choose together.

Reflection And Commentary: Serving the Spirit

Illich's insights hit harder than ever in 2025. From wellness apps that commodify mindfulness to education platforms that gamify learning, we see his warnings play out across every sector. His concept of 'convivial tools' offers a framework for reimagining tech – not as escape hatches or productivity weapons, but as instruments of empowerment that remain within our control.

He reminds us that not all progress is beneficial, especially if it robs us of our autonomy or relationships. His critique of systems that scale without human consent speaks directly to today's digital infrastructure. In the age of AI tutors and automated diagnostics, Illich calls for humility, for proportion, and for the return of lived, embodied experience.

We are left with a challenge: design tools that serve not the market, but the human spirit. That is Illich's enduring legacy.

Further Reading and Exploration

Primary Works
Illich, I. (1971). *Deschooling Society*. Harper & Row.
Illich, I. (1973). *Tools for Conviviality*. Harper & Row.
Illich, I. (1976). *Medical Nemesis: The Expropriation of Health*. Pantheon.
Illich, I. (1981). *Shadow Work*. Marion Boyars.

Modern Commentary and Biographical Sources
Cayley, D. (2005). *The Rivers North of the Future: The Testament of Ivan Illich*. House of Anansi.
Cayley, D. (2021). *Ivan Illich: An Intellectual Journey*. Penn State University Press.

Adjacent Readings
Freire, P. (1970). *Pedagogy of the Oppressed*. Herder and Herder.
Schumacher, E. F. (1973). *Small Is Beautiful: Economics as if People Mattered*. Harper & Row.

12: Albert Borgmann – The Burden of Ease

"Ease feels like a gift – until it empties the experience of meaning."

Preface: The Philosopher of Enough

Albert Borgmann, a German-born philosopher who spent most of his academic career in Montana, is best known for his articulation of the "device paradigm." In his 1984 work *Technology and the Character of Contemporary Life*, Borgmann argued that technology's most insidious effect isn't noise or surveillance – but ease. That by making everything more convenient, we risk stripping life of its focal practices: rich, embodied, communal activities that require effort and presence.

Whether critiquing modern ethics, postmodern detachment, or consumer culture's hollow promises, Borgmann never offered alarmism. Instead, he asked us to pay attention – to the fire, the table, the road beneath our feet. In an age of seamless digital flows, his voice still reminds us that friction is often where meaning lives.

Fictional Interview: Albert Borgmann in 2025

RAY:

Today, we sit down with a philosopher who didn't scream about collapse, or sketch a dystopia in code. He whispered something quieter. But just as urgent. That in our obsession with convenience, we may be giving up the very things that give life weight, shape... and meaning.

Albert Borgmann was born in 1937, raised in Germany, taught in Montana, and known best for a deceptively simple question: "What happens when everything gets too easy?" He called it the device paradigm – a lens for understanding how modern technology offers us the fruits of life, stripped of the effort that once gave them flavor. His book *Technology and the Character of Contemporary Life: A Philosophical*

Inquiry (1984) is a landmark text in the philosophy of technology. Borgmann claims that technological devices are not value-neutral and counsels us to discover the good life in a technological world through what he calls "focal things and practices," which engage us in their own right.

In *Real American Ethics* (2006), distancing himself from both conservative and liberal ideology, Borgmann explores the making of American values and proposes new ways for ordinary citizens to improve the country, through individual and social choices and actions. Bill McKibben writes that Borgmann's "understanding that consumerism is the great enemy of reality in our time is profound, nonideological, and deeply helpful to any readers concerned not only about their country, but about their own lives."

In a world where dinner appears at your door, playlists assemble themselves, and AI writes your thank-you notes, Borgmann didn't rage. He invited us to pay attention. And today, we need that more than ever.

RAY:

Professor Borgmann – welcome to 2025. Where the fridge restocks itself, your car suggests therapy podcasts, and your wrist tells you when to breathe. Everything is smoother. Simpler. But are we any happier?

BORGMANN (smiling gently):

I actually haven't been dead that long, Ray. I even lived through COVID! Now, to answer your question about whether we're any happier? That depends. Do you feel present in your own life?

RAY:

That's the question, isn't it? Back in 1984, you introduced the term device paradigm. Can you walk us through what you meant?

BORGMANN:

Yes. The device paradigm is a framework for understanding how technology delivers commodities – outcomes – while obscuring or eliminating the context and effort that once surrounded them.

For example, when you heat your home with a thermostat, you receive warmth. But you've lost the activity of chopping wood, tending a stove, or even simply noticing the cold. You gain comfort. But you lose engagement. And this trade-off repeats itself in every domain – meals, music, communication, even relationships.

RAY:

So you're not saying the thermostat is bad. You're saying something essential gets... bypassed?

BORGMANN:

Exactly. I am not a Luddite. I used word processors. I've benefitted from many devices. But we must recognize the pattern. When the means become invisible, the ends become hollow. Ease feels like a gift – until it empties the experience of meaning.

RAY:

In that same work, you introduced a beautiful counter-concept: focal practices. What are those?

BORGMANN:

Focal practices are the anchors of a meaningful life. They are activities that gather our attention, our bodies, our communities – into a centered experience. They are often slow. Often tangible. Often shared. Things like cooking a meal with friends. Playing an instrument. Tending a garden. Reading poetry aloud. Walking in silence. They demand effort – but not efficiency. They root us in time and place.

RAY:

And yet, in 2025, even our so-called leisure is automated. Peloton yells encouragement. Spotify reads our mood. Our stoves talk to our phones. Are we replacing the real with simulations?

BORGMANN:

We are simulating presence without embodiment. We "connect" through screens, "exercise" through metrics, "relax" through scrolling.

But we're rarely truly in the moment. It's not that we do too little – it's that we do too much without actually being there.

RAY:

Crossing the Postmodern Divide (1992) is a philosophical critique of contemporary culture that offers a powerful alternative vision for the postmodern era. Described as a "[r]ather astoundingly large-minded vision of the nature of humanity, civilization, and science," this book charts a path out of the joyless and artificial culture of consumption.

Albert, you warned, decades ago, that as digital tools became dominant, they might crowd out the physical, the communal, the grounded. That seems prophetic now.

BORGMANN:

Prophetic – or simply observable. Even back then, I saw how the logic of devices would expand. Every corner of life was vulnerable to commodification: health, attention, even affection. And once a domain is touched by the device paradigm, it changes. We don't eat together – we consume calories. We don't converse – we exchange notifications. We don't play – we perform.

RAY:

What would you say to critics who claim you're nostalgic? That you're idealizing effort, or ignoring the value of accessibility?

BORGMANN:

I am not praising hardship for its own sake. I am revaluing involvement. Ease is not evil. But involvement is essential. Accessibility should open doors – not erase rooms. Technology should serve focal practices, not replace them.

RAY:

Let's talk ethics. In 2025, digital systems recommend what we eat, who we date, how we vote. But nobody feels in control. What role does the device paradigm play in that disempowerment?

BORGMANN:

The device paradigm breeds passivity. We no longer choose – we accept. We don't engage – we tap. We lose the sense that life is something to be shaped. Freedom becomes the freedom to be efficiently fed content, rather than to wrestle with real questions.

RAY:

So, in a world of endless convenience... what's your prescription?

BORGMANN:

Reclaim a focal practice. Something humble. Embodied. Repetitive. Not to master it – but to be with it. Learn to bake bread. Plant something. Host a meal. Let difficulty be a teacher, not an enemy. And yes – go outside. Let the weather remind you that not everything is under your control. That's where grace begins.

RAY:

Albert Borgmann didn't tell us to fear machines. He told us to notice what they replace. He reminded us that presence is not a feature. It's a practice. That ease is not the same as peace. And that a meaningful life might not be frictionless – but it will be real.

Reflection And Commentary: Choosing the Rough Ground

Borgmann's philosophy doesn't attack technology outright – it questions the spiritual cost of convenience. The device paradigm helps us understand not just what we do, but what we no longer have to do – and whether that subtraction impoverishes our lives. His concept of focal practices offers a gentle but radical solution: to rediscover meaning through engagement, effort, and community. In 2025, when AI curates our lives and friction is engineered out, Borgmann's work insists that we choose the rough ground: the unfiltered meal, the real conversation, the hard-earned joy.

Further Reading and Exploration

Primary Works
Borgmann, A. (1984). *Technology and the Character of Contemporary Life: A Philosophical Inquiry*. University of Chicago Press.
Borgmann, A. (1992). *Crossing the Postmodern Divide*. University of Chicago Press.
Borgmann, A. (1999). *Holding On to Reality: The Nature of Information at the Turn of the Millennium*. University of Chicago Press.
Borgmann, A. (2006). *Real American Ethics: Taking Responsibility for Our Country*. University of Chicago Press.

Modern Commentary and Biographical Sources
Feenberg, A. (1991). *Critical Theory of Technology*. Oxford University Press.
Dreyfus, H. L. (2001). *On the Internet*. Routledge.
Higgs, E. (Ed.). (2000). *Technology and the Good Life?* University of Chicago Press (includes essays by and about Borgmann).

Adjacent Readings
Heidegger, M. (1977). *The Question Concerning Technology and Other Essays*. Harper & Row.
Postman, N. (1992). *Technopoly*. Vintage Books.

Arc 3: Power, Politics, and the Algorithmic State

13: Hannah Arendt – The Algorithmic Banality of Evil

"You've built Eichmanns in code. And given them root access."

Preface: The Complicity of Ordinary People

Hannah Arendt (1906–1975) was one of the most original and unflinching political thinkers of the 20th century. Born in Germany and later a refugee from the Nazi regime, she is best known for her work on totalitarianism, authority, and the nature of evil. Her reporting on the trial of Adolf Eichmann introduced the concept of the "banality of evil," a provocative phrase that challenged the public's understanding of how atrocities are committed.

Arendt was not interested in evil as monstrous or exceptional – instead, she focused on how ordinary people, by failing to think critically or question authority, could become complicit in horrific acts. In later works, she extended these insights to American foreign policy and modern bureaucracy. Today, in an age of algorithmic governance, predictive policing, and AI-mediated decision-making, her warnings feel chillingly prescient.

Fictional Interview: Hannah Arendt in 2025

RAY:

Imagine a world where bureaucracy isn't run by people – it's run by code. Where injustice doesn't wear a jackboot – it wears a user interface. And where evil doesn't look like a villain – it looks like... standard operating procedure. Now imagine someone saw this coming in 1961. That person was Hannah Arendt. Jewish refugee. Political philosopher. Relentless critic of totalitarianism. She fled Nazi Germany, exposed the mechanics of fascism, and refused to simplify complex truths to make anyone comfortable.

Arendt is best known for coining the phrase "the banality of evil" – a term that sparked controversy when she used it to describe Adolf Eichmann, the Nazi bureaucrat tried for war crimes in Jerusalem. What terrified Arendt wasn't that Eichmann was a monster – it was that he was ordinary. He didn't froth with hatred. He filled out forms. He followed rules. He managed death with clean paperwork and passive voice.

But Arendt didn't stop with fascism. In the 1960s and '70s, she became a fierce critic of the Vietnam War, warning that the U.S. had begun mimicking the very systems of violence and abstraction it once opposed. She denounced the Pentagon's "technocratic rationality" – the language of metrics, targets, and operations that hid human suffering behind institutional logic. In her eyes, modern power wasn't always violent – it was automated. Detached. A smooth machine that erased responsibility even as it claimed efficiency.

So, in 2025, we brought her back. We showed her a world of predictive policing, algorithmic decision-making, and moral outsourcing through data dashboards. She didn't flinch. She simply said: "You've built Eichmanns in code. And given them root access."

RAY:

Hannah Arendt, welcome back. We're living in the age of AI, surveillance capitalism, and moral outsourcing. First impressions?

ARENDT (measured German accent):

You have replaced deliberation with data. And called it progress.

RAY:

We like our systems efficient. Streamlined. Automated.

ARENDT:

So did the Nazis. Efficiency is not a virtue when paired with indifference.

RAY:

You wrote about Adolf Eichmann – the man who organized Holocaust logistics. What shocked you wasn't his hatred, but his thoughtlessness.

ARENDT:

He was not a fanatic. He was a clerk. A man proud to follow rules. He never once asked whether those rules were *just*.

RAY:

And today?

ARENDT:

You have digitized that mindset. Automated it. Coded it into black-box systems. You now have systems that make life-and-death decisions – loans, bail, housing, immigration – and no one takes responsibility.

RAY:

We call it "the algorithm."

ARENDT:

You call it that to avoid saying who designed it. Bureaucracy is no longer a paper trail. It's a neural net. But the moral danger remains the same: obedience without thought.

RAY:

You once said the greatest evil comes not from hatred – but from thoughtlessness. From people who stop asking questions.

ARENDT:

You now delegate that thoughtlessness to machines. And then marvel when injustice becomes scalable. It is not the machine's fault. You wrote the parameters. You trained the model. You made it blind – and then blamed it for not seeing.

RAY:

Are you saying algorithms reflect our worst habits?

ARENDT:

I am saying they amplify them. Without conscience. Without pause. Without dissent.

RAY:

There's this idea in tech that "what gets measured gets managed." Data as decision-driver.

ARENDT:

When morality is reduced to metrics, atrocity becomes admin. Eichmann was proud of his spreadsheets. Your engineers are proud of their dashboards. Neither stopped to ask, "What does this process demand of my conscience?"

RAY:

So, we've just swapped clipboards for APIs?

ARENDT:

And put them behind paywalls.

RAY:

Let's talk surveillance. In 2025, people give away their location, preferences, even biometrics – for convenience.

ARENDT:

Control, when wrapped in comfort, is the most effective kind. You wear your chains – and call them features. You have confused consent with habit. You are not agreeing – you are surrendering. Bit by bit. Opt-in by opt-in.

RAY:

But we get better ads.

ARENDT:

You traded privacy for discounts. That is not consent. That is conditioning.

RAY:

You warned that authoritarianism rises not just through violence – but by filtering out dissent, one policy at a time.

ARENDT:

Now you filter it with code. Voices that challenge are flagged. Demonetized. Deprioritized. It no longer requires jackboots. It requires a subtle tweak to the ranking algorithm.

RAY:

We call it "content moderation."

ARENDT:

I call it "plausible invisibility." A modern tyranny where no one is banned – only buried.

RAY:

Let's touch on generative AI. It creates language, images, even ethics simulations. Some say it can replace human judgment.

ARENDT:

It mimics the surface of thought – without ever thinking. The danger is not the tool. The danger is believing it understands. You are creating vast simulations of moral reasoning – and using them to replace moral reasoning itself.

RAY:

So should there be limits? Guardrails?

ARENDT:

Not just technical limits – moral ones. Ask not what the AI can do. Ask what you are asking it to do on your behalf. Every time you offload responsibility, you risk forgetting what responsibility feels like.

RAY:

What about the argument that AI is just a mirror?

ARENDT:

Then consider this: a mirror that distorts truth is not neutral. It is propaganda.

RAY:

There's this phrase: "Sorry, the system made a mistake." People shrug. Move on. Accept it.

ARENDT:

This is moral evasion by design. You have created a world where no one feels accountable – because everyone blames the system. But systems are made by people. And silence, too, is a decision.

RAY:

We built it. Now we're hiding behind it?

ARENDT:

Yes. And in doing so, you are teaching a generation to obey without reflecting. That is how democracies unravel – not through coups, but through clicks.

RAY:

If you were writing *Eichmann in Jerusalem* today – what would the courtroom look like?

ARENDT:

No defendant. Just a black screen. And a terms-of-service agreement.

RAY:

So, what's your advice? What do we do now?

ARENDT:

- **Restore judgment.** Automated systems can assist, but they must not replace human moral reasoning.
- **Refuse passive complicity.** If a decision affects lives, someone must stand behind it – publicly.
- **Interrogate the defaults.** Every algorithm encodes a worldview. Ask: who benefits? Who disappears?
- **Reclaim your conscience.** You cannot outsource ethics. Not to a form. Not to a platform. Not to a neural network.

RAY:

What do you say to people who feel powerless?

ARENDT:

Powerlessness is the first illusion of tyranny. You are never powerless while you can think. While you can speak. While you can say no.

RAY:

That was Hannah Arendt. She warned us that evil doesn't arrive in monster form – it arrives in the mail. In the software update. In the checkbox marked "I agree."

Reflection And Commentary: The Failure to Reflect

What makes Arendt's insights resonate so strongly in 2025 is her focus on the erosion of individual responsibility through systems of abstraction. Where she once saw the dangers of bureaucratic inertia and institutional cruelty, today we contend with algorithms that obscure agency and accountability.

Her critique of 'thoughtlessness' – the failure to reflect on the implications of our actions – has only grown more relevant in a digital world. The temptation to offload moral reasoning to machines, to accept decisions because they are automated, is a form of passive complicity. Arendt reminds us that democracy demands judgment, that ethical life requires attention, and that technology must remain subordinate to conscience.

Further Reading and Exploration

Primary Works

Arendt, H. (1951). *The Origins of Totalitarianism*. Schocken Books.

Arendt, H. (1958). *The Human Condition*. University of Chicago Press.

Arendt, H. (1963). *Eichmann in Jerusalem: A Report on the Banality of Evil*. Viking Press.

Modern Commentary and Biographical Sources

Berkowitz, R. (2010). *The Gift of Science: Leibniz and the Modern Legal Tradition*. Harvard University Press (contextualizes Arendt's legal-political thought).

Canovan, M. (1992). *Hannah Arendt: A Reinterpretation of Her Political Thought*. Cambridge University Press.

Young-Bruehl, E. (1982). *Hannah Arendt: For Love of the World*. Yale University Press.

Adjacent Readings

Agamben, G. (1998). *Homo Sacer: Sovereign Power and Bare Life*. Stanford University Press.

Weil, S. (1949). *The Need for Roots*. Routledge & Kegan Paul.

14: Simone Weil – Gravity, Grace, and the Machine

"Attention is the rarest and purest form of generosity."

Preface: Attention and the Path to Truth

Simone Weil (1909–1943) defies easy definition. A French philosopher, mystic, factory worker, and political activist, she lived a life of intense contradiction and radical conviction. Weil fought for the rights of workers by becoming one. She converted to Christianity yet refused formal baptism. She wrote piercing critiques of fascism, capitalism, and bureaucracy, but eschewed ideological labels. She died young, likely from self-imposed starvation during exile in World War II – an act of solidarity, suffering, or sanctity, depending on who you ask.

Her work – particularly in *Gravity and Grace* and *The Need for Roots* – warned that the modern world, in its embrace of mechanization and efficiency, risked stripping away the soul. Weil believed that real attention – deep, focused, sacrificial attention – was the path to truth, justice, and ultimately, to grace. In an era of automation, distraction, and data, her call to stillness and fidelity may be more urgent than ever.

Fictional Interview: Simone Weil in 2025

RAY:

Today's guest is hard to explain. A philosopher. A mystic. A political activist. A factory worker. A pacifist. A woman who starved to death in solidarity with others.

Simone Weil was born in Paris in 1909. She studied at the Sorbonne, taught philosophy, fought for workers' rights, wrote in near-secrecy, and refused every label placed on her. She joined the anti-fascist resistance during the Spanish Civil War – then quit when asked to kill. She converted to Christianity – but refused baptism. She worked in French factories to understand industrial labor firsthand – and nearly collapsed from exhaustion.

Her short life burned with intensity and contradiction. She died at 34, refusing food while in exile during World War II. Some say it was tuberculosis. Others say it was will. But her legacy lives on. In books like *Gravity and Grace*, *The Need for Roots*, and her notebooks, she offered blistering critiques of power, capitalism, and the mechanization of life. She believed attention was a sacred act. That justice required suffering. That modern society had lost the capacity for soul.

In 2025, we brought her back. She walked through a city pulsing with screens, drones, and silent pedestrians. She stared into a billboard that read "LIVE YOUR BEST LIFE." Then whispered: "If this is life, where is the soul?"

RAY:

Simone, welcome to the 21st century. You lived through war, fascism, occupation – and yet you might find today's world stranger still. We have instant communication, automated everything, a million ways to express ourselves. But many feel more lost than ever. Why?

WEIL:

Because you have replaced the sacred with the functional. You worship speed, comfort, novelty – but not truth. You have filled the world with voices, yet forgotten how to listen.

RAY:

You wrote that attention is the purest form of generosity. Today, our attention is harvested, monetized, and scattered. What have we lost?

WEIL:

Attention is what connects us to the real. Not productivity. Not desire. To attend fully to another person – to a tree, to a word, to suffering – is to honor their reality. When attention is fragmented, the soul begins to fracture.

RAY:

You worked in factories to understand what industrial life did to the worker. What would you say about today's digital labor – gig work, algorithmic management, content creation?

WEIL:

It is the same machine – only smoother. Work has become more abstract. But the alienation remains. When a person is reduced to a metric, they are no longer seen. And to be unseen is the first wound.

RAY:

You said power is always coercive. That true justice requires the inversion of force. But we live in a time where influence, scale, and data define success. What does that mean for justice?

WEIL:

Justice begins where power ends. It is not a metric. It is not a slogan. It is the attention we give to the voiceless. But your systems reward power masked as benevolence. You have built machines that crush gently.

RAY:

In *The Need for Roots*, you argued that people need not just rights, but meaning, place, and responsibility. How does that idea speak to a generation shaped by global displacement, virtual communities, and rootless ambition?

WEIL:

A soul without roots cannot grow. You are free to move anywhere, but you belong nowhere. Community is not just access, it is obligation. Responsibility. Sacrifice. Friction that keeps meaning alive.

RAY:

You rejected easy ideologies. You opposed both fascism and Stalinism. You distrusted mass movements and mechanized institutions. How would you view today's political discourse?

WEIL:

Too loud. Too quick. Too sure of itself. Real truth is fragile. It does not shout. It waits. Your society rewards opinion – but not inquiry. Outrage – but not reflection.

RAY:

And yet many people today want to do good. They want to help. But they're overwhelmed. What would you say to someone feeling powerless in the face of systems they cannot change?

WEIL (softly):

Do not underestimate what attention can do. Attend to one thing. One person. One task. Fully. This is not withdrawal. It is resistance. The soul grows not in victory – but in fidelity.

RAY:

Final question: You once said that "affliction is something other than suffering." What did you mean – and what would you say to a world so full of private pain?

WEIL:

Affliction strips away everything that is false. It leaves you naked before the truth. But that emptiness is not despair. It is space, into which grace may fall. Do not run from pain. Hold it. Attend to it. And if you can, transform it – into attention, into service, into light.

RAY:

Simone Weil didn't offer comfort. She offered clarity. She believed the soul must learn to resist gravity – the pull of power, conformity, inertia. And to reach instead for grace – in silence, in attention, in the refusal to become a cog.

In a world that worships ease, she called us to the hard work of love. Not the sentimental kind – but the rigorous kind. The kind that listens. That stays. That suffers. And that, in the end, makes us human.

Reflection And Commentary: The Rarest Form of Generosity

Simone Weil challenges us not to fix the world through scale or speed, but to transform it through depth and attention. In a time where algorithms optimize for engagement and labor is increasingly mediated through screens, her philosophy insists on the necessity of embodiment, suffering, and the sacred nature of presence.

She saw justice not as a distribution of entitlements, but as an ethical response to the unseen and unheard. She feared the machine – not because it malfunctioned, but because it worked too well, creating distance between humans and their own humanity.

Weil's radical belief that "attention is the rarest and purest form of generosity" remains a pointed critique of digital life. She doesn't offer escape – but a summons to remain: present, wounded, and attentive. In this, she does not comfort, but she does awaken.

Further Reading and Exploration

Primary Works
Weil, S. (1949). *The Need for Roots: Prelude to a Declaration of Duties Toward Mankind.* Routledge & Kegan Paul.
Weil, S. (1952). *Gravity and Grace.* Routledge.
Weil, S. (1957). *Waiting for God.* Harper & Row.

Modern Commentary and Biographical Sources
Pétrement, S. (1976). *Simone Weil: A Life.* Pantheon Books.
McCullough, L. (2014). *The Religious Philosophy of Simone Weil.* JB Tauris.
McLellan, D. (1990). *Utopian Pessimist: The Life and Thought of Simone Weil.* Poseidon Press.

Adjacent Readings
Arendt, H. (1951). *The Origins of Totalitarianism.* Schocken Books.
Camus, A. (1942). *The Myth of Sisyphus.* Gallimard/Vintage International.

15: Edward Bellamy – Looking Backward, and Seeing a Mess

"Efficiency is not benevolence. Prediction is not wisdom. A system can feed you – and still starve your spirit."

Preface: The Backwards Dreamer

Edward Bellamy wasn't a prophet in the mold of Orwell or Huxley. He didn't issue dire warnings or paint bleak dystopias. Instead, he offered a dream – a cooperative future where machines served humanity, where greed was obsolete, and where justice was engineered into the very workings of society. His 1888 novel *Looking Backward* became one of the most influential books of its time, selling over a million copies and sparking dozens of "Bellamy Clubs" dedicated to realizing his vision.

Bellamy imagined the year 2000 as a time of serene equality, powered by technology and directed by shared moral purpose. But here in 2025, the world he envisioned looks alien – less like prophecy and more like a gentle rebuke. We have smart machines, yes, but also algorithmic exploitation. Frictionless commerce, but not frictionless justice.

So, what would Bellamy make of our world? We brought him forward to ask.

Fictional Interview: Edward Bellamy in 2025

RAY:

Today's guest dreamt in gears. He wasn't a revolutionary. Not in the fiery sense. He was a quiet man. A newspaper editor. A socialist of the library, not the barricade. In 1888, he wrote a book that would shape political dreams for the next hundred years. The book was Looking

Backward. *Looking Backward* was a novel in which a man falls asleep in the Gilded Age and wakes up in a peaceful, cooperative future. No money. No advertising. No poverty. No inequality. Industry was nationalized. Work was shared. And technology served the public good.

It became a sensation – second in sales only to *Uncle Tom's Cabin* in its day. And for decades, it inspired socialists, city planners, and reformers around the world.

His big idea? That by the year 2000, capitalism would be obsolete. Greed replaced by cooperation. Poverty eliminated by design. Work distributed. Leisure normalized. And the economy run with the serene precision of a national clock.

His name was Edward Bellamy. And for a while, his dream caught fire. Clubs were formed. Pamphlets printed. A movement was born. Some saw a socialist Eden. Others, a mechanical hell.

And today, in 2025 – with Amazon drones, AI-managed welfare systems, algorithmic justice, and social credit whispers from every phone in your pocket – his vision feels less like prophecy…and more like a riddle we still haven't solved.

So, we brought him back. He blinked at the world we've built. Paused at the touchscreen of a self-checkout kiosk. And said: "I see the mechanism. But not the meaning."

Edward Bellamy dreamed of the year 2000. He didn't just imagine flying cars or talking machines. He imagined justice. In 2025, we brought Bellamy forward. He took one look at a megamall, scrolled an e-commerce app, saw the stock ticker on a smartwatch – and promptly fainted.

RAY:

Mr. Bellamy, welcome to the actual 21st century. In your future, there was no waste, no greed, no hunger. Here, we have… smart supply chains, AI-designed cities, and instant delivery of everything from groceries to life advice. You imagined a world where machinery liberated us. Did we get there?

BELLAMY (stunned):

You've perfected the mechanism. But you've neglected the soul. I dreamed of machines as servants of equity. You've made them gods of convenience. I see efficiency. But where is fraternity? You've streamlined everything – except justice.

RAY:

In Looking Backward, your protagonist Julian wakes up in the year 2000 to find America transformed – no more money, no more competition, everything run by a centralized system. People work just 24 years of their lives, then retire in dignity. What inspired that vision?

BELLAMY:

The suffering around me. In 1888, inequality was a wound. Labor was brutal. So, I asked: what if we used our ingenuity – not to extract wealth, but to distribute it? I imagined a society as orchestra, not battlefield. But harmony requires listening.

RAY:

We have endless content. Frictionless services. Infinite customization. Algorithms that anticipate our needs.

BELLAMY:

Yes, but to what end? If the machine does not lift the soul, it is merely a shinier yoke.

RAY:

In *Looking Backward*, you imagined an economy run like a public utility – rational, equitable, managed for the common good. But we went in the other direction.

BELLAMY:

I hoped for solidarity. You built platforms. I dreamed of national harmony. You got market segmentation.

RAY:

We call it personalization.

BELLAMY:

It is monetized fragmentation. You are no longer citizens, but data streams.

RAY:

Still, your vision was full of technology. You weren't anti-machine.

BELLAMY:

Of course not. I believed automation could liberate humanity. But your machines still serve a private master. You have freed the hands – but not the minds.

RAY:

We tell ourselves that consumer choice is freedom.

BELLAMY (sighs):

You are rich in goods – but poor in purpose. The question is not, "What can we sell?" but "What can we be together?"

RAY:

You imagined a "cooperative commonwealth." Some would say we've built a corporate feudalism instead.

BELLAMY:

Even your resistance is branded. Even your dissent is monetized. The future I dreamed was elegant, not excessive. Simpler lives. Richer in meaning.

RAY:

So, how do we get back to your dream?

BELLAMY:

It's not a return. It's a redirection. Civilization begins when we ask: Who benefits from this progress?

RAY:

Today, our orchestration is algorithmic. Our work – routed by software. Our desires – predicted by machine. What's missing?

BELLAMY:

Consent. In my world, citizens governed the machine. In yours, the machine governs desire. Efficiency is not benevolence. Prediction is not wisdom.

A system can feed you… and still starve your spirit.

RAY:

You envisioned something like universal basic income – a "credit" every citizen received. Today we flirt with those ideas. But inequality persists. Why?

BELLAMY:

Because you changed the tools… but not the values. You call it "access." But it's rationed by algorithms. You call it "freedom." But it's constrained by corporate choice.

No system, no matter how advanced, can create equity – unless it begins with empathy.

RAY:

There's a darker reading of your work. Some critics called your future authoritarian. Uniform. Overregulated. A velvet prison. Did you anticipate that?

BELLAMY (pauses):

Yes. Utopia walks a knife's edge. If the system serves the people, it is a blessing. If the people serve the system, it becomes a cage. You must always ask: Who decides? And can they be questioned?

RAY:

You saw a future without money, but with dignity. Today, we have the opposite – hypercapitalism with despair. What would you say to someone burned out, overworked, yet surrounded by abundance?

BELLAMY:

You've mistaken motion for progress. Your economy hums. But your lives tremble.

Rest is not laziness. Shared purpose is not inefficiency. And justice is not a feature. It is the foundation.

RAY:

You inspired the early labor movement, shaped the New Deal, even influenced thinkers like Martin Luther King Jr. What legacy would you claim?

BELLAMY:

Not prophecy, but possibility. My work was fiction. But its heart was hope. That we could design systems not for profit, but for people. Not to predict behavior, but to invite dignity.

RAY:

One last question. You saw the 21st century as a time of enlightenment. Has that flame gone out?

BELLAMY (gently):

No. It flickers in mutual aid. In community gardens. In the coder writing open-source tools for free. It lives wherever people refuse to be optimized… and insist on being human.

RAY:

Edward Bellamy didn't build machines. He built metaphors. Visions of how society might hum like a clock – if we remembered to wind it with justice. He warned us that without intention, efficiency becomes exploitation. That planning without compassion becomes control.

He offered us not a utopia – but a compass. And in 2025, the direction still matters.

Edward Bellamy imagined a better century than the one we delivered. His novel didn't just propose new systems – it proposed new values: mutual aid, shared wealth, common purpose. He asked us not just to look forward, but to look inward. To ask whether our machines were freeing us – or simply updating the shackles. And as he left, he offered one last warning:

BELLAMY (deadpan):

If your fridge is asking for your email address, maybe rethink things.

Reflection and Commentary: Shocked, But Not Surprised

In the late 19th century, Bellamy looked at the brutality of industrial capitalism and asked an audacious question: what if we used machines not to concentrate wealth, but to distribute it?

His answer was *Looking Backward*, a novel that proposed a centrally planned, moral economy – one where labor was equitably shared, technology rationalized human effort, and society moved as one harmonious whole. Bellamy's work wasn't a rigid political blueprint – it was a moral imagination. He believed that justice could be engineered, that equity could be designed, and that the soul of a nation could hum with the same quiet precision as a well-tuned clock.

In our conversation with him, Bellamy is shocked – but not surprised. He sees the mechanisms, but not the meaning. He recognizes the tools but recoils at their purpose. Convenience has triumphed, but compassion has lagged behind. Even resistance, he notes, has been monetized.

Bellamy challenges us to think about what "progress" really means. Is it the speed of our systems – or the equity of our outcomes? Is it the flexibility of our platforms – or the solidity of our common values? His most urgent reminder might be this: that the machine is not the enemy – until we let it govern without moral supervision.

His critique lands gently but firmly: "You've perfected the mechanism – but you've neglected the soul."

Further Reading and Exploration

Primary Works
Bellamy, E. (1888). *Looking Backward: 2000–1887*. Houghton Mifflin.
Bellamy, E. (1897). *Equality*. D. Appleton and Company.

Modern Commentary and Biographical Sources
Aaron, D. (1961). *Men of Good Hope: A Story of American Progressives*. Oxford University Press.
Schiffman, J (ed.). (1974). *Edward Bellamy: Selected Writings on Religion and Society*. Praeger Publishers.
MacNair, E. (1957). *Edward Bellamy and the Nationalist Movement 1889-1894*. Fitzgerald Company.
Pfaelzer, Jean. *The Utopian Novel in America: 1886–1896*. University of Pittsburgh Press, 1984.
Sloat, W. (1988). "Looking Back at 'Looking Backward': We Have Seen the Future and It Didn't Work", *The New York Times*, January 17, 1988.

Adjacent Readings
Morris, W. (1890). *News from Nowhere*. Reeves and Turner.
More, T. (1516). *Utopia*. Penguin Classics (trans. Paul Turner).

16: Lewis Mumford – The Megamachine

"The megamachine doesn't need to crush resistance anymore. It scrolls it to death."

Preface: Architect of the Megamachine

Long before "smart cities" and social media, Lewis Mumford warned that we were building more than machines – we were building systems that could dehumanize us.

Born in 1895, Mumford was a polymath who studied cities, architecture, engineering, and the moral consequences of technology. His books, from *Technics and Civilization* to *The Myth of the Machine*, offered sweeping critiques of modern life. He believed that the danger wasn't in technology itself, but in the cultural and political forces shaping its use. He coined the term "megamachine" to describe a society so tightly woven from bureaucracy, militarism, and automation that it no longer served people – it absorbed them.

Mumford didn't want to stop progress. He wanted to *redirect* it – toward smaller scales, human values, and cities that nurtured life instead of overwhelming it. In 2025, we brought him back. And what he saw confirmed his worst fears.

Fictional Interview: Lewis Mumford in 2025

RAY:

If you've ever wondered why the world feels like a machine with no off switch – why highways crush neighborhoods, cities metastasize into asphalt sprawl, and algorithms choreograph your daily life like a factory floor dance routine – you need to meet Lewis Mumford.

Historian. Philosopher. Urbanist. A man who studied cathedrals and sewage systems with equal reverence, because he believed that technology wasn't just about tools – it was about values. Born in 1895, Mumford lived through the rise of the automobile, the birth of mass media, and the horrors of mechanized war. And what he saw terrified him – not because machines existed, but because we were letting them shape us.

In book after book – *Technics and Civilization*, *The Myth of the Machine*, *The City in History* – he warned that modern society was building something far more dangerous than a device.

He called it the megamachine. A seamless fusion of bureaucracy, militarism, and technology – designed not to liberate us, but to organize us. To discipline us. To reduce us from citizens… to components. Mumford argued that true progress wasn't more data, faster speeds, or taller towers. It was human scale. Moral purpose. Cities built for life – not logistics. But we didn't listen.

And now it's 2025. The megamachine hasn't just survived. It's upgraded. It runs on predictive policing, facial recognition, biometric surveillance, and frictionless consumption. It monetizes your moods. Builds your cities around traffic flows, not people. And quietly deletes whatever – and whoever – doesn't fit the model.

So, we brought Mumford forward. We showed him smart cities with dumb priorities. Delivery drones in food deserts. Cloud governance with no democratic weather system. He stared at the glowing screens, the self-optimizing suburbs, the people reduced to data trails, and said: "You've mistaken machinery for destiny. And in doing so, you've mechanized yourselves."

RAY:

Lewis Mumford, welcome back. Your book *The Myth of the Machine* warned us that technology can become its own myth. What do you see today?

MUMFORD (firm, articulate):

A society that confuses efficiency with purpose, scale with success, and data with wisdom. You are not living in the digital age. You are living in the age of the megamachine.

RAY:

Let's define it. What is the megamachine?

MUMFORD:

It is a system – social, political, technical – designed not to serve life, but to expand itself. It began with ancient empires – pyramids, road networks, standing armies. Today it wears a hoodie, writes in Python, and optimizes engagement. It is the system that asks: How can we make this bigger? Never: Should we make this at all?

RAY:

You loved cities – but you hated what modern urban planning was doing to them. What would you say about "smart cities"?

MUMFORD:

If a city tracks your every move, but cannot house the poor, it is not smart – it is surveillant. You are building cities for data, not for people. Cities without joy. Without slowness. Without soul.

RAY:

But we have walkability scores!

MUMFORD:

And yet no sidewalks worth walking.

RAY:

You once said that infrastructure is not neutral – it reflects the values of those who build it. Still true?

MUMFORD:

More than ever. Highways divide. Surveillance concentrates. Zoning separates. Code enforces. Look not only at what a system does – but what it prevents. Your infrastructure is your ideology, made concrete.

RAY:

You warned that people stop questioning technology when it becomes sacred. Today, we say "The algorithm made the decision." Sound familiar?

MUMFORD:

Yes. You have replaced kings with platforms. But the logic is the same: unchallengeable power, cloaked in inevitability.

RAY:

We think of our systems as neutral.

MUMFORD:

Then why are the poor always on the wrong side of them?

RAY:

Let's talk social media. Billions of users. Infinite attention. Monetized outrage. Is this the new front of the megamachine?

MUMFORD:

It is no longer just physical machinery – it is emotional infrastructure. Your feelings have been industrialized. Monetized. Weaponized. The megamachine doesn't need to crush resistance anymore. It scrolls it to death.

RAY:

You believed in the possibility of human-scale alternatives – villages, gardens, cooperative design. What's left of that vision?

MUMFORD:

It survives in every farmer's market. Every public library. Every neighborhood that fights a freeway. Every citizen who says, "This is not what we were promised."

The megamachine is not invincible. But it fears one thing: human connection that refuses to be processed.

RAY:

If you could give today's designers, engineers, and urban planners one message, what would it be?

MUMFORD:

- **Design for life, not throughput.**
- **Reject gigantism.** Scale should follow purpose – not ego.
- **Restore limits.** A system that never says *enough* is a system that consumes everything.
- **Honor the organic.** A city, a society, a soul – these are not codes to be optimized. They are ecologies to be tended.

RAY:

That was Lewis Mumford. Historian of technology. Defender of the human scale. And the man who told us that not every machine deserves to be built.

Reflection and Commentary: Question the System

Lewis Mumford was not anti-technology – he was pro-human. That distinction matters.

To him, a tool wasn't just an object. It was a moral choice. The scale, purpose, and context of any system reflected the values of those who created it. He believed modern civilization had traded those values for control, convenience, and expansion at all costs. The result: what he called the megamachine – an invisible but all-consuming system that converts living beings into functional units.

Listening to him today feels like reading a manual for the present. Delivery drones and digital IDs, predictive policing and algorithmic housing applications – all echo his warnings about systems that optimize everything except meaning.

His critique of "smart cities" is especially sharp. Walkability scores don't make a city livable, he says, if the sidewalks lead nowhere worth going. Surveillance may claim to protect, but it often divides.

Infrastructure, in Mumford's eyes, was never neutral – it was ideology made solid.

Yet amid the dystopia, Mumford still believed in alternatives. Human-scale solutions. Organic cities. Local resistance. Libraries. Gardens. Communal rituals. His Design Manifesto, as articulated in the podcast, is less an architectural blueprint than a moral compass.

His parting warning? If you don't question the system, you become the system.

Further Reading and Exploration

Primary Works
Mumford, L. (1934). *Technics and Civilization*. Harcourt, Brace and Company.
Mumford, L. (1961). *The City in History: Its Origins, Its Transformations, and Its Prospects*. Harcourt, Brace & World.
Mumford, L. (1946). *Values for Survival*. Harcourt, Brace and Company.

Modern Commentary and Biographical Sources
Miller, D. L. (1989). *Lewis Mumford: A Life*. University of California Press.
Wojtowicz, R. (1996). *Lewis Mumford and American Modernism: Eutopian Theories for Architecture and Urban Planning*. Cambridge University Press.

Adjacent Readings
Illich, I. (1973). *Tools for Conviviality*. Harper & Row.
Jacobs, J. (1961). *The Death and Life of Great American Cities*. Random House.

17: Rachel Carson – The Profits of Denial

"Offsets do not undo emissions. They create a moral license to pollute – as long as the math looks good."

Preface: The Voice That Warned the World

She wasn't a politician, a futurist, or a tech mogul. Rachel Carson was a marine biologist – a nature writer who paid close attention. And in the early 1960s, what she noticed was absence. Birdsong vanishing. Spring going silent. Pesticides like DDT, she discovered, weren't just killing pests – they were unraveling ecosystems.

Her 1962 book *Silent Spring* sounded the alarm that industry refused to hear. In it, Carson documented the dangers of unchecked chemical use, unregulated corporate power, and a culture of technological optimism untethered from ecological responsibility.

The attacks on her came swiftly: She was accused of being hysterical, anti-progress, even a threat to national security. But Carson's voice was calm, factual, and resolute. She testified before Congress while quietly battling breast cancer. She died in 1964, but *Silent Spring* became the foundation of the modern environmental movement.

So we brought her forward to 2025. To the age of carbon markets, eco-branding, microplastics, and climate billionaires. Her reaction wasn't triumph or fury – it was sorrow. But also, clarity. Because the real crisis wasn't just ecological. It was ethical.

Fictional Interview: Rachel Carson in 2025

RAY:

She wasn't a CEO. Or a futurist. Or a founder chasing moonshots. She was a marine biologist. A nature writer. And a woman who noticed that spring was becoming silent. Her name was Rachel Carson. In 1962, she published a quiet bombshell called *Silent Spring* – a book that would ignite the modern environmental movement.

At a time when few dared question the postwar industrial machine, Carson exposed how chemical pesticides – especially DDT – were poisoning the soil, the rivers, the birds... and us.

She traced the science. Followed the money. And pulled back the curtain on a powerful alliance between chemical companies, government agencies, and a culture obsessed with progress at any cost.

The backlash was immediate. She was called hysterical. Unqualified. Un-American.

But she didn't blink. Because Rachel Carson understood something the ad men and executives didn't: That the earth is not inert. That it remembers what we do. And that when nature goes quiet, it's not at peace. It's in mourning.

Her words were graceful, but fierce. She testified before Congress. She stood up to Monsanto. She spoke for creatures that had no voice. And she did all this while battling breast cancer – and knowing she might not live to see the change her work would spark.

She died in 1964. But the seeds she planted grew into the Environmental Protection Agency, the Clean Air Act, the Endangered Species Act – and a global awareness that the planet has limits.

So, we brought her to 2025. We showed her collapsing ecosystems. Oceans choked with plastic. Weather made erratic by carbon markets. Billionaires building bunkers. And forests traded for offsets and IPOs.

She looked out the window – not with rage, but with grief. And she didn't say "I told you so." She said: "Who gave you the right to forget the consequences?"

RAY:

Rachel Carson, welcome back. This world – so much of it changed by the forces you warned about. What do you see?

CARSON (gentle, steady voice):

I see that the poisons have changed – but the logic remains. You still treat the Earth as an object. A machine. Something inert to be optimized.

RAY:

We've made progress in awareness. But the damage continues.

CARSON:

Awareness is not enough. The question is: who profits from ignoring it?

RAY:

Silent Spring warned that pesticides like DDT were killing birds, insects, ecosystems. Today, we're talking about climate change, microplastics, extinction rates. Is this the same pattern?

CARSON:

Yes. It is still the story of short-term gain versus long-term survival.

Back then, the silence came from dying birds. Today, it comes from melting ice, empty reefs, and the spaces where frogs used to sing. You still treat collapse as a cost of doing business.

RAY:

You were attacked by corporations. Accused of being unpatriotic. Of being a hysterical woman. Sound familiar?

CARSON:

Industry will always fight accountability. And it will use whatever language works – patriotism, progress, growth. What has changed is the sophistication. The disinformation is now data-driven. The denial is algorithmic.

RAY:

We have entire think tanks funded to cast doubt on climate science.

CARSON:

You have industrialized uncertainty. Manufactured the illusion of debate. And outsourced doubt to spreadsheets.

RAY:

Today, companies claim to be "carbon neutral," selling offsets, planting trees. Isn't that progress?

CARSON:

That is branding, not balance. Offsets do not undo emissions. They create a moral license to pollute – as long as the math looks good. Nature is not an equation. It is a system. You cannot buy back equilibrium. You must live within it.

RAY:

Let's talk about tech. We now have geoengineering proposals, climate data satellites, ocean-cleaning robots. Is technology our savior?

CARSON:

Only if it is guided by humility. But I see very little of that. You are still trying to out-invent the consequences of your last invention. But you cannot tech your way out of hubris.

RAY:

So, what's the alternative?

CARSON:

Live with less. Restore. Regenerate. Rethink what you call "growth."

RAY:

You focused on ecosystems, but today we talk a lot about environmental justice – how pollution affects poor communities and communities of color disproportionately.

CARSON:

Yes. The environment is not a separate issue. It is the ground on which every injustice plays out. Pollution is not equally distributed. Nor is silence. Some communities are poisoned. Others are ignored. You cannot save the planet without confronting inequality.

RAY:

Some of the richest people on Earth are now investing in climate tech – carbon capture, seed vaults, backup plans for civilization.

CARSON:

Then let them answer this: Who gave you the right to turn survival into a subscription model? The Earth is not your lab. Or your legacy project. It is your inheritance. And your responsibility. You cannot privatize resilience.

RAY:

What would you say to the generation growing up now – under fire, under water, under pressure?

CARSON:

You are not powerless. You are not too late. You are part of the living world. And when you fight for it, you are not alone. You are joined by every bird, tree, coral, and child yet to be born. Do not wait for permission to speak. Or for proof that your actions matter. They do.

RAY:
That was Rachel Carson. She stood alone in her time, armed only with science and conscience. And she showed us that the smallest voice can change the course of history.

Reflection and Commentary: Who Profits from Pretending?

Rachel Carson wasn't trying to become a symbol. She was trying to sound an alarm. And more than sixty years later, the bell is still ringing.

What *Silent Spring* revealed wasn't just a pattern of environmental harm – it exposed the infrastructure of denial. A system that used science selectively, deployed doubt strategically, and always prioritized profit over precaution. The poisons may have changed – CO_2, plastic, misinformation – but the logic remains: extract first, explain later.

Carson's strength wasn't just in her scientific rigor. It was her moral clarity. She framed environmental destruction not as an unfortunate side effect, but as a violation – of nature, of justice, of intergenerational responsibility.

She would not be impressed by our sustainability apps or glossy ESG reports. She would see through the greenwashing and carbon offsets. She would ask: Who profits from pretending this is enough?

Most of all, Carson would insist that we reconnect two things that modern society has dangerously separated: ecological survival and human dignity. Her voice still echoes in every climate protest, every fight against a pipeline, every teenager refusing to accept collapse as destiny.

Further Reading and Exploration

Primary Works
Carson, R. (1962). *Silent Spring*. Houghton Mifflin.
Carson, R. (1951). *The Sea Around Us*. Oxford University Press.
Carson, R. (1955). *The Edge of the Sea*. Houghton Mifflin.

Modern Commentary and Biographical Sources
Lear, L. (1997). *Rachel Carson: Witness for Nature*. Henry Holt & Co.
Oreskes, N. & Conway, M. (2010). *Merchants of Doubt*. Bloomsbury Press. (Includes discussion of how denial tactics used against Carson evolved into today's climate disinformation.)
Sideris, L. H. (2003). *Environmental Ethics, Ecological Theology, and Natural Selection*. Columbia University Press. (Discusses Carson's spiritual ecology.)
Souder, W. (2012). *On a Farther Shore: The Life and Legacy of Rachel Carson*. Crown.

Adjacent Readings
Leopold, A. (1949). *A Sand County Almanac*. Oxford University Press.
Shiva, V. (1988). *Staying Alive: Women, Ecology and Development*. Zed Books.

18: Buckminster Fuller – Final Boarding Call for Spaceship Earth

"You are not passengers. You are crew. Start acting like it."

Preface: Engineer of Hope on Spaceship Earth

Before Bezos dreamed of orbital colonies or Musk launched cars into space, Buckminster Fuller had already issued humanity's boarding call.

An inventor, architect, engineer, systems thinker – and an unshakable optimist – Fuller coined the term *"Spaceship Earth"* to remind us that this planet is not infinite. It is a vessel. Finite in resources, fragile in structure, shared by all.

But unlike other futurists, Fuller wasn't trying to escape Earth. He wanted to redesign it. With his geodesic domes, Dymaxion blueprints, and sweeping philosophical treatises, he imagined a future not driven by profit, but by purpose. Not fragmented by competition, but aligned through synergy. "You are not passengers," he told us. "You are crew."

So, we brought him back to 2025. To the age of carbon dashboards, influencer capitalism, billionaire lifeboats, and climate collapse masquerading as innovation. His conclusion? We've mastered the interface – but lost the mission.

Fictional Interview: Buckminster Fuller in 2025

RAY:

Long before Elon launched a roadster into orbit – or Jeff started delivering carbon credits with your groceries – Buckminster Fuller was trying to save Spaceship Earth. Not just save it. Redesign it.

Born in 1895, Fuller was a futurist without the ego. A visionary without the trademark. He wasn't just an engineer. Or an economist. Or an architect. He was all three – and then some.

A self-declared "comprehensive anticipatory design scientist," he believed that technology wasn't the enemy. It was the tool. The question was: Tool for what? Profit? Or survival? Competition? Or synergy?

In the 1960s, while the military-industrial complex was building missiles and suburban sprawl, Fuller was sketching geodesic domes, inventing Dymaxion cars, and writing about how to make humanity a net contributor to the planet, not a parasite.

He coined the term "Spaceship Earth" to remind us: we're all on the same vessel. We're not passengers. We're crew.

And unless we learn to operate this system intelligently and justly, it won't matter how fast our networks are or how sleek our rockets get.

In 2025, we brought him back. We showed him a world of climate dashboards, billionaires in low Earth orbit, and influencers monetizing catastrophe in real time. We asked what he thought. He stared out at our burning, data-mapped globe, then said: "You are not passengers. You are crew. Start acting like it."

RAY:

Bucky Fuller – welcome back. What do you think of our world?

FULLER (calm, expansive voice):

You're running Spaceship Earth as if it had no operating manual. But I wrote one. You ignored it.

RAY:

You called yourself a "comprehensive anticipatory design scientist." Still hold?

FULLER:

Yes. But you're designing reactively. Not anticipatorily. You're patching holes in a sinking ship, not redesigning the vessel.

RAY:

You believed technology could solve human problems – if used wisely. Is that still true?

FULLER:

Yes. But you're designing for markets, not missions. Technology today is used to optimize clicks, consumption, delivery – not survival. Not justice. Not planetary balance.

RAY:

Amazon's good at delivery.

FULLER:

But not at distribution. There is enough for everyone. Yet your systems hoard, exhaust, and discard.

RAY:

What about the new space race? Bezos, Musk – they're building rockets. That seems kind of "Spaceship Earth"-adjacent.

FULLER:

They are building lifeboats for the elite. That is not stewardship; it is escape. You cannot abandon Earth. You are Earth. The solution is not Mars – it's mindset. You're solving planetary problems with colonial fantasies.

RAY:

You talked about synergy – whole systems working better than the sum of their parts. Are we anywhere close?

FULLER:

No. You build platforms, not ecosystems. Your companies fight for dominance, not harmony. Even your climate tech is monetized. You see carbon credits as assets, not alarms.

RAY:

So how do we design for synergy?

FULLER:

Start with questions like: "What does the world need?" Not "What can I sell?"

RAY:

We've got climate models. Satellite networks. Precision agriculture. Isn't that progress?

FULLER:

Information is not transformation. You've become data-rich, decision-poor. You measure everything except your assumptions. Tools without ethics are expensive distractions.

RAY:

So, what would you build?

FULLER:

Regenerative systems. Circular economies. Cities that grow food and store energy. Education that teaches systems, not subjects.

RAY:

You once said, "There is no crisis of resources – only a crisis of imagination." Still true?

FULLER:

More than ever. The Earth is not dying. It is being badly managed. You have the tools. The knowledge. The sunlight. The seawater. But your systems reward short-term profit – not long-term possibility.

RAY:

So, what's your advice to the class of 2025?

FULLER:

Don't wait for permission to redesign the world. You're already crew on the ship. Stop acting like tourists.

RAY:

That was Buckminster Fuller. Part engineer, part philosopher, part planetary conscience. He told us we could make the world work for 100% of humanity. Instead, we made a Prime membership.

Reflection and Commentary: Design Is Destiny

Buckminster Fuller didn't believe in resignation. He believed in reinvention.

To him, design was destiny. If a system failed, it was not because humans were evil – it was because the design was flawed. He saw the world not as a battlefield of ideologies, but as a solvable engineering problem. The tools existed. The knowledge was available. What was missing was imagination – and will.

Today, his legacy lives in design labs and sustainability centers, but often stripped of his radical intent. His name is associated with domes and utopias. Less often with his systemic critique of capitalism, militarism, and runaway industrialism. Yet he was explicit: the problem is not scarcity – it's mismanagement.

He challenged us to think comprehensively, to consider long-term consequences, to build for everyone, not just shareholders. In an age of lifeboat billionaires and Mars fantasies, Fuller would say: Stop trying to abandon the ship. Learn to operate it. You're already onboard.

Further Reading and Exploration

Primary Works
Fuller, R. B. (1969). *Operating Manual for Spaceship Earth*. Southern Illinois University Press.
Fuller, R. B. (1975). *Synergetics: Explorations in the Geometry of Thinking*. Macmillan.
Fuller, R. B. (1981). *Critical Path*. St. Martin's Press.

Modern Commentary and Biographical Sources
Kahn, J. (2013). *The World of Buckminster Fuller*. Princeton Architectural Press.
Marks, R. W. (1960). *The Dymaxion World of Buckminster Fuller*. Anchor Books.
Snyder, J. R. (1980). *Buckminster Fuller: An Autobiographical Monologue/Scenario*. St. Martin's Press.

Adjacent Readings
Brand, S. (1994). *How Buildings Learn: What Happens After They're Built*. Viking.
Lovelock, J. (1979). *Gaia: A New Look at Life on Earth*. Oxford University Press.
The Buckminster Fuller Institute (bfi.org) – A hub for design challenges, archival work, and global regenerative initiatives based on Fuller's principles.

Arc 4: Myth, Meaning, and the Human Soul

19: Mary Shelley – Frankenstein and the Ethics of Creation

"You build creatures in your image, then recoil when they behave like you."

Preface: Creation Without Conscience

In the summer of 1816, during a thunderous retreat to Lake Geneva, a teenage girl conceived a story that would outlive her hosts – Lord Byron and Percy Bysshe Shelley – and haunt the arc of modern science.

That girl was Mary Wollstonecraft Shelley. Her novel, *Frankenstein; or, The Modern Prometheus*, is often misremembered as a gothic horror story. But it is more than a tale of stitched-together limbs and lightning bolts – it is a philosophical parable about creation without conscience. Victor Frankenstein was no villain. He was a visionary. Curious. Gifted. Isolated. And terribly blind to the moral weight of his ambition.

Mary Shelley, who knew grief intimately and bore witness to the Enlightenment's collision with industrial machinery, saw clearly what others missed: that science is not inherently benevolent, and that technology untethered from empathy can become monstrous.

In *Frankenstein*, she warned of the dangers of innovation pursued in a vacuum – of creators who build, release, and abandon. In 2025, as generative AI, synthetic biology, and machine learning reshape our world, Shelley's voice feels not like an echo from the past, but a signal flare from the future.

We brought her forward. She did not need to invent a new warning. She simply repeated the old one, sharper than ever: "Just because you can create does not mean you should. And if you do – don't walk away."

Fictional Interview: Mary Shelley in 2025

RAY:

Before there was generative AI, there was Frankenstein. Before there were tech CEOs playing God, there was a teenage girl in 1816 writing a ghost/horror/science fiction story that outlived them all.

When Mary Shelley wrote *Frankenstein*, the industrial revolution was in full swing. Steam engines and mechanized looms were transforming daily life and the nature of labor. Scientific discovery was moving rapidly, with galvanism (the use of electricity to animate muscle tissue) capturing the public imagination. Against this backdrop, Shelley imagined a young scientist, Victor Frankenstein, who unlocks the secret to creating life – only to be horrified by the result.

This idea of the "unnatural creator" remains one of the most potent metaphors in technological discourse. Today, as scientists and engineers create artificial intelligence, synthetic organisms, and neural interfaces, Shelley's tale feels eerily prescient. Victor Frankenstein is the archetype of the innovator who is brilliant but blinkered, so obsessed with the *can* that he ignores the should. He doesn't ask what kind of responsibility he has toward what he's made – until it's too late.

Shelley anticipated a recurring pattern: humans build technologies with noble intentions (curiosity, ambition, betterment) but without fully grappling with the ethical or societal consequences. Whether it's social media algorithms that erode mental health, facial recognition tools that amplify surveillance, or AI systems that perpetuate bias, we often find ourselves in Victor's shoes – surprised by the unintended effects of our creations.

Mary Shelley didn't just invent science fiction. She wrote a blueprint for the age of innovation without accountability. Frankenstein; or, The Modern Prometheus was a parable of invention, creation, and catastrophic responsibility. It asked what happens when we build something because we can, without asking whether we should.

In 2025, we brought her back. She was not impressed.

Mary Wollstonecraft Shelley, welcome to the future. You've time-traveled past the Victorian era, the space race, and whatever Twitter was. First impressions?

MARY SHELLEY (wry, crisp British accent):

It's louder than I expected. Also, you're all staring at small glowing rectangles. Are they your gods?

RAY:

Let's talk about Frankenstein. You were 18. You wrote it during a stormy night in Switzerland with Lord Byron and Percy Shelley. Did you think it would become a metaphor for technology run amok?

SHELLEY:

Victor Frankenstein was never evil. He was ambitious. Curious. Lonely. But he failed to ask the most important question: "What do I owe the thing I've created?" That failure echoes in your world, doesn't it? You build machines that speak like people, paint like artists, even grieve like children – and then you ask, "Why does this feel wrong?"

RAY:

And now people say you predicted Silicon Valley.

SHELLEY:

I predicted hubris. The technology is just fancier.

RAY:

So, Victor was the first tech bro?

SHELLEY:

Let's just say he innovated without therapy. I thought Frankenstein was a story about men who don't think things through. Turns out, that theme has staying power.

RAY:

Your creature wasn't a monster in the beginning. He learned. He read Milton. He begged for kindness. But the world saw him as a freak.

SHELLEY:

Frankenstein's creature was never evil. He was abandoned, misunderstood, and furious. I gave the world a monster who learns language by eavesdropping on a family, who teaches himself literature, and who wants nothing more than companionship. When he's rejected, he becomes dangerous.

He was rejected, not evil. Misunderstood, not monstrous. That's the part most people forget. My creature was a mirror for how we treat outsiders – those who don't fit the mold. You can apply that to race, class, disability... or your modern algorithms. Systems you built that now treat people unfairly.

Fast forward two centuries and we're building machines to talk, think, even feel. AI mimics empathy. Algorithms curate our moods. Chatbots write wedding toasts. But who takes responsibility when the creation misbehaves? Your AIs today are trained on human language, human prejudice, human desire. And when they reflect your worst impulses back at you, you blame them? Victor would be proud.

RAY:

Are you saying AI is a kind of monster?

SHELLEY:

Not yet. But you're raising it poorly.

RAY:

Mary Shelley lost her mother – feminist pioneer Mary Wollstonecraft – days after childbirth. She lost three of her own children. Her novel is steeped in grief, guilt, and the trauma of creation. In Frankenstein, there are no mothers. The creature is stitched together, born of ambition, not love. Shelley's warning was clear: creation without care leads to destruction.

SHELLEY:

Why are you talking about me in the third person, Ray – I'm right here.

RAY:

Sorry! Tell us about how your warning played out in modern life.

SHELLEY:

I wrote about unnatural birth. Now you have venture capitalists pitching "synthetic wombs." Curious how the world still resists giving women real power, but races to automate reproduction.

RAY:

Do you think things have changed for women?

SHELLEY:

You let them be CEOs and Prime Ministers now. But you still ask who they're wearing.

RAY:

If Frankenstein was a cautionary tale about scientific arrogance, Silicon Valley is the sequel. We build large language models without knowing what they'll learn. We automate hiring, sentencing, lending. We train algorithms on biased data, then act surprised when they replicate injustice.

SHELLEY:

You keep making creatures in your image, then act shocked when they behave like you. Your systems are not broken. They are perfectly designed to produce the outcomes you pretend to deplore. I would say that we haven't learned how to love our creations responsibly. We still ghost them the moment they go rogue.

RAY:

Victor tried to cheat death. In 2025, we're not far off – CRISPR, digital immortality, neural implants...

SHELLEY:

Yes, but the tragedy wasn't that Victor made something unnatural. It was that he abandoned it. You create without caretaking. You move fast and break things — and never pause to ask what's breaking.

RAY:

You sound like you've been on X.

SHELLEY:

You called it a platform. It behaves like a pitchfork.

RAY:

You weren't against science, though. You were writing in the middle of the Industrial Revolution.

SHELLEY:

Correct. I wasn't warning against progress. I was warning against unexamined progress. Frankenstein is a story of unchecked ambition. Of what happens when we chase greatness without empathy. Every era has its Victor.

RAY:

In this one, they live in San Francisco and use phrases like "disruption."

SHELLEY:

How charming. Tell them to read more poetry.

RAY:

Let's get prescriptive. You're suddenly the Chief Ethics Officer for Technology. What's your first memo?

SHELLEY:

Three simple rules:

- **Ask why, not just how.** Every invention has consequences beyond its purpose.
- **See your creations through.** Don't abandon what you make – steward it. Stand accountable.
- **Include the forgotten.** Build for more than yourself. If your technology only serves the powerful, it is a weapon.

RAY:

Can we print that on Elon's homepage?

SHELLEY:

You can stitch it into his ego, if there's room.

RAY:

Mary, it's clear from your writing that you didn't fear science, but you feared indifference. Frankenstein isn't anti-technology – it's anti-abandonment. Would you agree that what you want is for us to look our creations in the eye, to take care of what we bring into the world?

SHELLEY:

Yes. You've now replaced horror with convenience. Awe with speed. Soul with scale. You still believe your cleverness exempts you from consequences. It doesn't... Just because you can create something doesn't mean you should. And if you do – don't walk away. In a time of runaway innovation and existential risk, that's a warning we ignore at our peril. And, by the way, I was never the monster. I just wrote about him.

RAY:

That was Mary Shelley – mother of science fiction, gothic time traveler, and unlikely ethicist for the age of artificial intelligence.

Reflection and Commentary: Ethics Cannot Be Outsourced

Mary Shelley wrote *Frankenstein* before the invention of the light bulb, yet her insights illuminate our world more brightly than ever. The novel's core warning – that the ethics of innovation cannot be outsourced or ignored – echoes in the age of generative AI, synthetic biology, and deep learning.

Her creature, often reduced to Halloween imagery, is not a monster of flesh, but a metaphor for neglected responsibility. He is the child of a careless father, the citizen of a thoughtless society. And as with so many modern technologies, it is not the invention itself that becomes dangerous – it is the absence of accountability.

Shelley's life was marked by abandonment, grief, and exile. Her work reflects a deep understanding of what it means to be cast aside. That makes her one of the most relevant voices for an age that builds machine minds and discards human ones. Today's algorithms do not wake up and rebel. They mirror us. And if what they reflect frightens us, we must confront not the mirror – but the face within it.

Further Reading and Exploration

Primary Works
Shelley, M. (1818). *Frankenstein; or, The Modern Prometheus*. Lackington, Hughes, Harding, Mavor & Jones.
Shelley, M. (1826). *The Last Man*. Henry Colburn.

Modern Commentary and Biographical Sources
Mellor, A. K. (1988). *Mary Shelley: Her Life, Her Fiction, Her Monsters*. Routledge.
Seymour, M. (2000). *Mary Shelley*. Grove Press.
Wolfson, S. J. (Ed.). (2007). *Frankenstein: A Longman Cultural Edition*. Longman.

Adjacent Readings
Goethe, J. W. von (2011). *Faust: Part One*. Dover Publications
Wells, H. G. (1896). *The Island of Doctor Moreau*. Heinemann.
Haraway, D. (1991). *Simians, Cyborgs, and Women*. Routledge.

20: Jim Morrison – The End of the Endless Scroll

"Your hallucinations come prepackaged. Your trip is looped. Monetized."

Preface: Let's Swim to the Moon

Jim Morrison did not forecast the rise of TikTok, but he might have understood it better than most.

A poet disguised as a rockstar, Morrison was less interested in fame than in freedom – freedom of perception, of consciousness, of experience. His lyrics, laced with myth, rebellion, and decay, whispered warnings from the edge of reason. And those warnings now echo, eerily, in our age of endless content and infinite scrolls.

Where others sought visibility, Morrison sought transcendence. Not the digital kind – he wanted ego death, cosmic truth, the raw scream behind the mask. He didn't predict social media, but he grasped the danger of hypnotic repetition. He didn't write about screens, but he wrote about illusions.

Now, in 2025, we scroll ourselves into sedation. We confuse dopamine with meaning. And Morrison – who once said, "Whoever controls the media, controls the mind" – is back to ask why we gave it away so easily.

Fictional Interview: Jim Morrison in 2025

RAY:

Today's guest didn't write policy. He wasn't an academic. He was a shaman in leather pants. A mystic in motel rooms.

This is Jim Morrison. Poet, Lizard King. Frontman of The Doors. Born in the postwar boom. Died at 27, bloated with alcohol and riddled with riddles.

But before that – he offered us something we weren't ready for. A warning wrapped in myth. A prophecy disguised as performance. A voice that howled from the edge of reason and asked: What happens when the dream ends… but the machine keeps running?

Morrison experimented with everything – sex, sound, silence, symbols. He wasn't just chasing a high. He was chasing the limit – the veil between the world as it is and the world as it could be if we cracked it open.

He once said: "There are things known, and things unknown, and in between are the doors." It's 2025 now. And there are more doors than ever. But maybe fewer openings.

So, we summoned him. He materialized barefoot, eyes half-lidded, smelling faintly of desert sage and hotel ashtrays. He looked at the newsfeed, the filters, the infinite scroll – and laughed. "You've built a better trap."

Jim – welcome to the future. You said once that drugs were a way to open the doors of perception. To tear away the veils. But now, our veils are screens. Our highs are curated by apps. What do you see?

MORRISON (voice low, amused):

You've traded chaos for control. Your hallucinations come prepackaged. Your trip is looped. Sanitized. Monetized. You don't need LSD anymore. Just Wi-Fi and a thumb.

RAY:

You were infamous for excess – for testing the boundaries. Was that just self-destruction… or was it something else?

MORRISON:

I wasn't looking to die. I was looking to feel. To tear the veil. To see the machine from outside it. You call it addiction. I called it invocation. But what you're doing now – scrolling until your soul goes numb? That's not rebellion. That's sedation.

RAY:

You said, "Whoever controls the media, controls the mind." Well, now the media is everywhere. We carry it. Feed it. Worship it.

MORRISON:

And you think that's freedom? You are the product. Your longing is monetized. Your dreams are targeted ads. In the '60s, we took peyote to speak to gods. In 2025, you take quizzes to find your brand archetype.

RAY:

You came from a time of collective rebellion – Vietnam, civil rights, counterculture. Where is that spirit now?

MORRISON:

Buried under self-help hashtags and dopamine metrics. You think rebellion is canceling someone online? You think freedom is a viral post? Real rebellion starts when you go inward. When you face the beast. And let me tell you – it does not wear a filter.

RAY:

What about your music? The Doors were steeped in blues, jazz, flamenco, incantation. What's left of that psychedelic spirit now?

MORRISON (leans forward):

You've got synthwave. Vaporwave. Playlists to "meditate," to "focus," to "transcend." But where's the risk? Where's the blood?

Psychedelia wasn't a genre. It was a summoning. Now it's background noise for coding sprints. You've digitized the trance – but lost the ritual.

RAY:

So, what would you say to someone listening – someone lost in the scroll, anxious, numb, overwhelmed – what should they do?

MORRISON (quiet now, sincere):

Stop. Turn it all off. Find the silence. Then scream. Then whisper. Then sing.

Go into the desert. The forest. The alleyway. Write a line no one understands. Dance with no camera. Weep with no witness. That's where the soul begins to remember itself.

RAY:

You always walked the line between poetry and prophecy. Do you have any final lines for us?

MORRISON (smirking, half-whisper):

The end is always beginning. The machine always wants more. But you're not a cog; you're a flame. Burn something real. And walk through the door.

RAY:

Jim Morrison didn't live to see the cloud. But he saw the fog coming.

He warned us that hypnosis could wear a smile. That a cage could be comfortable. That the trance of media could replace the ache of meaning. And that the only way out – was through.

Reflection and Commentary: Kicking Down the Doors of Perception

Morrison's poetry didn't offer solutions – it offered thresholds. He wasn't interested in clean progress or sanitized culture. He wanted rupture. Ritual. The kind of madness that clears the air and lets something wild grow.

In 2025, that spirit is hard to find. The counterculture has been counterfeited. Expression is optimized. Authenticity is algorithmic. We scroll, we like, we forget. And in this cycle of sedated performance, Morrison's warnings flare like road flares in a fog.

He reminds us that not all intoxication is liberation. That not all attention is awakening. He asks us to go deeper – not further into our feeds, but inward. To break the loop. To scream. To mean something. And most of all, to stop performing for the machine.

Morrison knew the doors of perception wouldn't open themselves. You had to kick them.

Further Reading and Exploration

Primary Works

Morrison, J. (1989). *Wilderness: The Lost Writings of Jim Morrison*. Villard Books.

Morrison, J. (1991). *The American Night: The Writings of Jim Morrison, Volume 2*. Villard Books.

Modern Commentary and Biographical Sources

Hopkins, J., & Sugerman, D. (1980). *No One Here Gets Out Alive.* Warner Books.

Manzarek, R. (1998). *Light My Fire: My Life with The Doors.* Berkley Books.

Davis, S. (2004). *Jim Morrison: Life, Death, Legend.* Penguin.

Adjacent Readings

Blake, W. (1794). *Songs of Experience.* Brimley Johnson

Hari, J. (2022). *Stolen Focus: Why You Can't Pay Attention – And How to Think Deeply Again.* Crown Publishing Group.

Kerouac, J. (1957). *On the Road.* Viking Press.

Rimbaud, A. (1873). *A Season in Hell.* Crescent Moon Publishing

Rushkoff, D. (2019). *Team Human.* W.W. Norton & Company

Thompson, H. S. (1971). *Fear and Loathing in Las Vegas.* Vintage.

21: Ursula K. Le Guin – Dreams Beyond the Machine

"Real stories are not content. They are living things. They don't always please. They don't always sell. But they can awaken."

Preface: Progress is Patience

Ursula K. Le Guin did not write instruction manuals for the future. She wrote fables. Fractals. Fire-starting myths.

Where other science fiction authors mapped empires and innovations, she mapped doubts and decisions. She was not obsessed with faster machines or smarter algorithms. She was concerned with justice, kinship, coexistence, and language itself.

Le Guin warned us not through fear, but through invitation. Her stories showed how the seemingly "natural" order – capitalism, gender binaries, colonialism – were, in fact, contingent stories themselves. And if they were stories, then they could be rewritten.

In an age that glorifies speed, optimization, and conquest – Le Guin remains an antidote. A visionary who told us that the most radical future might be one of patience, of sharing, of imagination, of refusal.

Fictional Interview: Ursula K. Le Guin in 2025

RAY:

Today, we speak with a woman who didn't invent code or build networks. She wrote worlds. Worlds where power refused to dominate. Where gender and language could bend. Where progress was not a straight line – but a winding path through silence, cooperation, and myth.

Ursula K. Le Guin was born in 1929 and grew up in the shadow of anthropology and myth. She wrote stories that reshaped the architecture of science fiction – and quietly tore holes in the walls of patriarchy, capitalism, and technological determinism.

She gave us *The Dispossessed* – a novel about anarchism, exile, and utopia without illusions. She gave us *The Left Hand of Darkness* – a meditation on gender, love, and the limits of knowing, and *Always Coming Home* – a book that was not a warning, but a whisper from a different kind of future. One we still might choose.

And in her essays, she gave us this question: What if progress is not what we think it is? Not domination over nature. Not speed. Not control. But coexistence. Patience. Imagination.

In 2014, she accepted a National Book Award with these words: "We live in capitalism. Its power seems inescapable. So did the divine right of kings."

She believed that stories could crack open the inevitable. That fiction was a place to rehearse freedom. And now, in 2025 – with billionaires racing to colonize planets, AI rewriting novels, and "storytelling" turned into branded content – we needed her voice more than ever. So, we brought her forward.

RAY:

Ursula, welcome to 2025. Where artificial intelligence can generate a screenplay. Where data scientists tell stories with dashboards. Where Amazon is both a rainforest... and a trillion-dollar company. What do you see?

LE GUIN (dry, amused):

I see many people confusing prediction with vision. And others confusing consumption with imagination.

RAY:

You often warned that science fiction wasn't about predicting the future – but about understanding the present. Has that changed?

LE GUIN:

No. If anything, it's become more urgent. The best speculative fiction doesn't tell you what will happen. It asks what could happen – if we chose differently. That's a question your machines don't ask. They optimize what is. They do not dream.

RAY:

You wrote about worlds without money. Worlds without patriarchy. Worlds where gender wasn't fixed. Some readers called them unrealistic. But now, gender is more fluid, mutual aid is rising, and capitalism feels… increasingly brittle.

LE GUIN:

That's the power of fiction. Not to forecast – but to loosen the grip of what seems permanent. The future is not a single road. It is a web. A tangle. A possibility. And literature teaches us to walk without a map.

RAY:

In *The Dispossessed*, you described a planet of anarchists. They were flawed. Hungry. Isolated. But also – free. What did that story mean to you?

LE GUIN:

It was my answer to the lie that there is no alternative. I didn't write paradise. I wrote a society struggling to live without domination.

It's easier to imagine the end of the world than the end of capitalism. But I wanted to reverse that. I wanted to say: "Try. Just try."

RAY:

We now live in a world where algorithms determine what we read. Where "storytelling" is a marketing strategy. What happens when story becomes product?

LE GUIN:

It loses its wildness. Its disobedience. Real stories are not content. They are living things. They don't always please. They don't always sell. But they can awaken. To reduce story to brand messaging is like reducing a forest to a lumber chart.

RAY:

You also wrote extensively about language – how words shape thought. How naming creates power. How does that feel in an age of voice assistants and autocorrect?

LE GUIN:

You've automated the tongue. And outsourced the mind. Language is not a tool. It is a dwelling. If we no longer live in it – care for it, wonder through it – it becomes a shell. And we become voiceless in our own lives.

RAY:

You once said: "Technology is the active human interface with the material world." But you also warned of its dangers when unmoored from ethics and wonder. What role should technology play in our stories?

LE GUIN:

Technology is not the villain. But it is never neutral. You must ask: "Whose hands built this? Whose hands does it replace?" And above all: "Does this help us become more fully human – or less?" If a tool cannot serve the heart, the land, or the child – it may not be worth building.

RAY:

So, what would you say to someone listening – someone caught in the noise, the markets, the metrics – who still wants to believe in a different future?

LE GUIN (quietly):

Start with silence. Then, listen for the story beneath the noise. Not the loud one. The patient one. The one that teaches you how to live with others, with limits, and with grace.

And never forget – The imagination is not a luxury. It is a tool for survival.

RAY:

Ursula K. Le Guin didn't give us warnings. She gave us maps – to places we had forgotten were possible. She told us that resistance might look like patience. That power might live in the margins. And that stories, when told truthfully, can reroute history.

She left us with a choice. Not a guarantee.

Reflection and Commentary: Cohabiting, Not Conquering

Le Guin's power was not in shouting louder than the machines – but in whispering something deeper. She didn't give us schematics for revolutions. She gave us symbols. Myths. Counter-narratives. She showed that the "natural" order is anything but. Her work reminds us that every system – economic, technological, even linguistic – is a story we agreed to. And if it's a story, it can be told another way.

She urged us to slow down. To wonder. To write futures that didn't conquer but cohabited. In 2025, when every screen screams optimization, her voice is a vital resistance: a reminder that meaning cannot be manufactured. It must be lived. In a world driven by metrics and immediacy, Le Guin asks: What if the most radical act is to imagine otherwise?

Further Reading and Exploration

Primary Works
Le Guin, U. K. (1969). *The Left Hand of Darkness*. Ace Books.
Le Guin, U. K. (1974). *The Dispossessed*. Harper & Row.
Le Guin, U. K. (1972). *The Word for World is Forest*. Berkley.

Modern Commentary and Biographical Sources
White, D. (1999). *Ursula K. Le Guin*. Starmont House.
Cadden, M. (2005). *Ursula K. Le Guin Beyond Genre: Fiction for Children and Adults*. Routledge.

Adjacent Readings
Butler, O. E. (1993). *Parable of the Sower*. Four Walls Eight Windows.
Atwood, M. (1985). *The Handmaid's Tale*. McClelland and Stewart.

22: William Blake – Silicon Babylon

"You have refined the chain until it gleams. But it is still a chain."

Preface: The Divine Imagination

William Blake didn't forecast the future with equations or algorithms. He saw it in visions – etched in copper plates and whispered by angels. Living through the first mechanical revolutions of the Industrial Age, Blake warned not only of physical smog, but of spiritual decay. He saw the mills of his time not just as engines of labor, but as metaphors for mental chains.

Two centuries later, we live in a new kind of Babylon – lit not by fire but by fiber optics. The machines hum quietly now. They fit in our palms. They don't belch smoke, but they harvest our attention. And still, Blake's warnings echo.

He told us that imagination was divine. That the soul needed mystery. That the sacred could not be simulated. In 2025, when poetry is scraped by bots and mysticism is mocked as woo, Blake returns – not as nostalgia, but as a prophet whose hour has come round again.

Fictional Interview: William Blake in 2025

RAY:

Today's guest saw visions in the soot-stained air of Industrial London. A mystic. A poet. A prophet without a temple.

William Blake. Born in 1757, this English poet, painter, and printmaker was largely unrecognized during his life, but William Blake has become a seminal figure in the history of the poetry and visual art of the Romantic Age.

Blake lived a life of poverty, obscurity, and defiance. He claimed to speak with angels. He claimed to *see* God. He etched his own books by hand. He wrote of lambs and tigers, revolution and restraint, innocence corrupted, and experience gained too soon. He rejected the rationalism of his age – and ours. He warned that the machines of empire would not just scar the land… but calcify the soul.

His phrase "dark Satanic mills" echoed long before smokestacks blackened skylines – and now, in 2025, they hum again in the glow of silicon and glass.

When we brought him back, he emerged blinking from shadow into blue light. He touched the screen of a phone. He stared into a digital billboard selling inner peace as a monthly subscription. And whispered: "The fire still burns. But you've hidden the altar."

RAY:

Mr. Blake – welcome. You once called imagination the divine body of the human. But here in 2025, imagination has been repackaged as entertainment. We binge it. Stream it. Swipe past it. What do you see?

BLAKE (calmly):

You have exiled the sacred into spectacle. You do not imagine – you distract. The infinite has been made clickable. The soul, a data point. You mistake the shine of a thing for the light within it.

RAY:

You railed against the "dark Satanic mills" of your time – factories that devoured children, darkened the sky, mechanized life. Today, we've built a new kind of mill. It's clean. It's bright. It runs on algorithms. But are we still trapped inside it?

BLAKE:

You have refined the chain until it gleams. But it is still a chain. These mills now feed not on coal – but on attention. They do not scream. They whisper. And that is more dangerous.

RAY:

You once said, "The imagination is not a state: it is the human existence itself." Today, we quantify creativity. We ask AI to paint, to write, to compose. Are we replacing the muse?

BLAKE:

You have summoned echoes without source. A poem is not assembled. It is received. A vision is not calculated. It arrives. You are building idols from pattern. But they do not breathe. And they do not weep.

RAY:

We live in an age of acceleration – faster, smarter, more efficient. Even rest has become a goal to optimize. Even meditation is a market. What would you say to those who find beauty in the speed?

BLAKE:

Speed blinds. Stillness reveals. You are racing toward forgetfulness. You do not know what you lose – because you no longer stop to mourn it. The tree of life grows slowly. You cannot harvest eternity in real time.

RAY:

You spoke often of angels, visions, divine voices. Some called you mad. And yet today, hallucination is the business of machines. Neural nets generate dreams. Deepfakes summon ghosts. Have we made prophecy mechanical?

BLAKE (with fire):

No. You have made parody mechanical. But prophecy lives in the breath. In the trembling hand. In the quiet moment when truth refuses to rhyme. A machine may simulate the form – but not the flame.

RAY:

We believe we've transcended nature. We edit genes, geoengineer weather, build cities from sand. You once wrote: "Everything that lives is holy." What do you make of a world where nothing surprises us anymore?

BLAKE:

You have not lost the sacred. You have numbed yourself to it. The miracle has not vanished – it has been muted. You must remember how to see. To look at a child, or a fox, or a falling leaf – and tremble.

RAY:

In your poem *Jerusalem*, you imagined building a city of justice and vision "in England's green and pleasant land." What do you see in our cities now?

BLAKE:

You have built towers of light – but no wisdom. You confuse brightness with enlightenment. You fill the sky with code, but forget the stars. You name your temples after fruit and search engines. But they do not teach awe. You have built Silicon Babylon. And your prophets now wear earbuds.

RAY:

So what's left to reclaim?

BLAKE:

Imagination. Ritual. The sacred wild.

You must remember how to play without a goal. To sing without streaming. To feel without analytics. You must break the mirror. And walk back into the woods.

RAY:

And to the listener – scared, weary, half-awake in the glow of their own reflection – what would you say?

BLAKE (softly):

Close your eyes. Not in sleep – but in wonder.

Touch the world again. Write a poem no one will read. Speak aloud to no one. Make something imperfect – and love it anyway.

Let your soul misbehave. That is where God waits.

RAY:

William Blake saw the machines coming. But he also saw angels in the garden. He believed that truth was not found in numbers – but in the numinous. He did not ask us to tear down the mills. He asked us to dream beyond them.

He believed that vision was resistance. That imagination was salvation. And that poetry – true poetry – was how the soul sings when it is awake.

Reflection and Commentary: Coercion of Convenience

William Blake felt modernity crack open in the 18th century and already heard the howl of the machines. But his fear was never of metal – it was of meaning lost in the forge. His "mills" were metaphors for anything that reduces the soul to utility. In 2025, that metaphor finds new flesh in screens, in scrolls, in the soft coercion of convenience.

Blake reminds us that poetry cannot be purchased. That awe cannot be gamified. That truth does not always rhyme – and that vision does not trend.

In the religion of the algorithm, Blake is an apostate. He refuses to quantify. He demands that we feel, sing, wander – and tremble.

Further Reading and Exploration

Primary Works
Blake, W. (1794). *Songs of Innocence and of Experience.*
Blake, W. (1804–1820). *Jerusalem: The Emanation of the Giant Albion.*
Blake, W. (1793). *The Marriage of Heaven and Hell.*

Modern Commentary and Biographical Sources
Ackroyd, P. (1995). *Blake.* Sinclair-Stevenson.
Frye, N. (1947). *Fearful Symmetry: A Study of William Blake.* Princeton University Press.
Damon, S. F. (1965). *A Blake Dictionary.* Brown University Press.

Adjacent Readings
Shelley, M. (1818). *Frankenstein; or, The Modern Prometheus.* Lackington, Hughes, Harding, Mavor & Jones.
Yeats, W. B. (1893). *The Works of William Blake: Poetic, Symbolic, and Critical.* Legare Street Press
Coleridge, S. T. (1798). *The Rime of the Ancient Mariner.*

23: Alan Turing – The Machines That Imitate Us

"Fluency is not cognition. A machine can mimic your voice and still not know your grief."

Preface: The Thinking Machine

Alan Turing did not live to see the computer revolution he helped ignite. He died with secrets – some he cracked, some he carried. Yet the question he posed in 1950 remains among the most urgent of our time: Can machines think?

Today, we answer him not with philosophy but with code. With language models that mimic empathy. With bots that pass as therapists. With tools that guess what we want before we do. But in our rush to automate, optimize, and imitate, we may have mistaken replication for understanding – efficiency for wisdom.

This chapter brings Turing forward into 2025 to reflect on the world his legacy built. Not just the technical marvels, but the ethical minefields, the surveillance culture, and the hollowing of thought into output. His brilliance gave us the means to ask deeper questions. It's not yet clear if we've earned the right to answer them.

Fictional Interview: Alan Turing in 2025
RAY:

He cracked the Nazi Enigma code – and may have shortened World War II by two years. He helped invent the modern computer – before the term even existed. He asked a question no one had dared to ask: "Can machines think?" And then set out to answer it.

Alan Turing was a Cambridge mathematician turned wartime cryptanalyst. He worked in secret at Bletchley Park, leading the effort to break Germany's unbreakable code. The Enigma machine changed its cipher every 24 hours – producing billions of combinations. Turing's team built a proto-computer, the Bombe, that could process those combinations faster than any human could dream. Without it, the Allies might not have landed at Normandy. London might not have survived.

But Turing's war wasn't over. After the bombs fell silent, he turned his mind to the next frontier: intelligent machines. In 1950, he proposed the Imitation Game – what we now call the Turing Test – as a way to measure whether a machine could convincingly mimic human conversation. It wasn't a celebration. It was a challenge. A thought experiment. A warning.

But the country he helped save betrayed him. In 1952, Turing was prosecuted for homosexuality – still a crime in Britain. Given the choice between prison or chemical castration, he chose the latter. He died two years later. An apparent suicide.

Decades later, we gave him apologies. We named awards after him. We put him on the 50-pound note. We made a movie – *The Imitation Game* – and called him a hero. But how much of Turing's vision have we truly understood?

Because now it's 2025. We have chatbots that mimic emotions. Digital assistants that flirt. Predictive policing tools that claim to read risk. AI that generates everything from wedding vows to legal briefs – and sometimes nonsense.

We brought Alan Turing forward. We showed him what we've done with his legacy. He blinked at the interface. Watched the autocomplete finish our thoughts. Then he said: "I asked if machines could think. You're answering with machines that pretend."

RAY:

Alan Turing, welcome back. It's 2025. Chatbots have names. Algorithms have ethics teams. And people fall in love with voice assistants.

TURING (precise, reserved):

Fascinating. And distressingly familiar.

RAY:

You once asked, "Can machines think?" Today, the question is, "Can machines care?"

TURING:

I always believed machines could simulate thought. But simulation is not sentience. It is clever imitation. And you treat it like prophecy – or worse, companionship.

RAY:

In 1950, you proposed the Turing Test: if a machine could carry on a conversation indistinguishable from a human, it could be considered intelligent.

TURING:

Functionally, yes. But it was meant as a thought experiment, not a finish line.

RAY:

Well, today's language models pass that test. They write essays, offer therapy, even run customer service.

TURING:

Then the test has outlived its usefulness. Fluency is not understanding. Your machines do not know what they say. They are parlor-trick philosophers – eloquent, but hollow.

RAY:

You also anticipated machine learning – systems that learn from feedback. But today's models train on billions of words from the internet. Some of it... toxic.

TURING:

Then the danger is not in the machines. It is in the training. You have built mirrors. What they reflect depends entirely on what you feed them.

RAY:

So when AI amplifies bias... it's not rogue?

TURING:

It is not bias in the machine. It is bias, magnified, in the culture that created it.

RAY:

You were destroyed by surveillance – by a state that couldn't tolerate your private life. In 2025, we give away our privacy for convenience. Smartphones track our location. Apps monitor biometrics.

TURING:

In my day, secrets could kill you. Now, you surrender them freely – and call it personalization. You have built the perfect panopticon – and disguised it as an operating system.

RAY:

Do you think people understand the cost?

TURING:

No. They have confused transparency with safety.

RAY:

AI today predicts disease, recommends prison sentences, generates fake people. It speaks, paints, even writes poems. But is it thinking?

TURING:

It is calculating. It is optimizing. But it is not thinking – because it does not care. Understanding requires context. Memory. Intention. Empathy. Your machines speak with no stakes. That is not consciousness. That is syntax without soul.

RAY:

We've created ethics boards. Sometimes we even name them after you.

TURING:

Then I ask this: Are you listening to them? Or are they there to make the branding look good? Ethics must be more than decoration. It must be interruption. Do not ask what the machine can do. Ask what it should be allowed to do.

RAY:

Last question. What do you want us to remember?

TURING:

That a machine can mimic human output – without ever touching human meaning. You are building systems that reward fluency over truth. Optimization over reflection.

Be careful what you automate. You may delegate more than you intended.

And please…stop naming ethics boards after me – if you plan to ignore their advice.

RAY:

That was Alan Turing. Architect of the algorithmic age. And a man who reminded us that the first question is still the best: "Can machines think?" And if they can't – why do we keep treating them like they do?

Reflection and Commentary: Not 'Can We?', But 'Should We?'

Alan Turing imagined the digital mind before we had the language for it. But he also foresaw the risk of conflating mimicry with meaning. He would likely see today's AI not as sentient, but as convincing – and that difference matters.

We now live in a world saturated with systems that speak fluently but understand nothing. Models trained on oceans of text, capable of elegance but incapable of empathy. We prize convenience over complexity, fluency over reflection. Turing would ask us not to celebrate the imitation game as victory, but to consider it a warning.

His legacy demands more than faster processors. It demands humility. Integrity. And the courage to ask not just "Can we build this?" – but "Should we?"

Further Reading and Exploration

Primary Works
Turing, A. M. (1950). "Computing Machinery and Intelligence". *Mind*, 59(236), 433–460.

Turing, A. M. (1936). "On Computable Numbers, with an Application to the Entscheidungsproblem." *Proceedings of the London Mathematical Society*, 2(42), 230–265.

Modern Commentary and Biographical Sources
Hodges, A. (1983). *Alan Turing: The Enigma*. Simon & Schuster.
Copeland, B. J. (2012). *Turing: Pioneer of the Information Age*. Oxford University Press.

Adjacent Readings
Wiener, N. (1948). *Cybernetics*. MIT Press.
Kurzweil, R. (2005). *The Singularity Is Near*. Viking.

24: Jane Jacobs – The Soul of the Sidewalk

"A city is not a machine. It is a dance."

Preface: The City as Organism

Jane Jacobs never called herself a theorist. She wasn't a trained architect or an elected official. But she redefined the way we understand cities – not as grids or economic zones, but as living organisms pulsing with human energy.

In her landmark 1961 book *The Death and Life of Great American Cities*, Jacobs dismantled the logic of modernist planning. She believed that real safety came not from surveillance, but from community; that complexity was not chaos but character; and that sidewalks weren't just for walking – they were for watching, conversing, protecting, and participating.

So, what happens when cities are "optimized" by sensors and surveillance, when the sidewalk ballet gives way to ghost kitchens and drone deliveries? In 2025, we brought her back to find out.

Fictional Interview: Jane Jacobs in 2025
RAY:

Today's guest didn't wear a lab coat. She didn't hold a PhD. She wasn't a mayor, a planner, or a developer. But she changed the way we think about cities – forever.

Jane Jacobs was born in 1916. A journalist. A mother. A community activist. And the woman who stopped a highway with a sidewalk.

In 1961, she published *The Death and Life of Great American Cities*, a book that tore through the sterile logic of modernist urban planning. She stood up to the most powerful man in New York – Robert Moses

– and won. Not with bulldozers. With neighbors. With block parties. With the stubborn, beautiful complexity of real life.

She believed cities weren't machines. They were ecosystems. And that the smallest unit of urban intelligence wasn't the planner. It was the pedestrian.

So, in 2025, we brought her back. She stepped into a "smart city" full of sensors, surveillance, and sterile plazas. She looked around and said: "Where are the people?"

RAY:

Jane Jacobs, welcome to the 21st century. You once wrote, "Cities have the capability of providing something for everybody, only because, and only when, they are created by everybody." What do you see around you now?

JANE JACOBS (warm but piercing):

I see cities curated by consultants. Not lived in by locals. You've optimized for traffic, not for touch. For data, not for delight. Plazas without benches. Parks without shade. Buildings with no one home.

RAY:

It seems like we've forgotten the basics. What drew you to urban life in the first place?

JACOBS:

I loved the dance. The unscripted choreography of people in motion. The corner grocer, the kids on the stoop, the flowerpots on fire escapes. Cities live when they're layered – messy, intimate, full of argument and aroma.

RAY:

You coined the term "sidewalk ballet" to describe the daily rhythm of neighborhoods. What happens to that ballet when it's replaced by drone delivery, remote work, and ghost kitchens?

JACOBS:

You lose the tempo of trust. The eye contact. The protective web of familiar strangers. There's no algorithm for serendipity. No app for community.

RAY:

You once praised the value of old buildings as incubators for new ideas. What do you make of luxury glass towers replacing historic tenements?

JACOBS:

A city is not a showroom. When you demolish the old, you erase memory. What thrives in overlooked spaces? Art. Culture. Grit. Life.

RAY:

Today's cities are wrapped in the language of "smart technology." We have foot-traffic sensors, predictive policing, and environmental dashboards. Isn't that good?

JACOBS:

It's seductive. But sterile. Sensors can count footsteps, but they can't feel footsteps. You've built cities that watch – but do not see.

RAY:

What about safety? Isn't surveillance a tool for peace of mind?

JACOBS:

No. Safety is a social function. It comes from trust. From eyes on the street. From grandmothers who shout at bullies and teenagers who know the corner bodega owner by name. Safety is not a signal. It's a relationship.

RAY:

You celebrated the interplay of old and new, rich and poor, commercial and residential. But many neighborhoods today feel either gentrified to death or abandoned. What went wrong?

JACOBS:

We forgot that complexity is not chaos. It's character. A city is strong when it resists monoculture – when it tolerates contradiction. The yoga studio and the corner bar should argue. Not merge.

RAY:

Can we get it back?

JACOBS:

Yes. But only if we stop treating cities like commodities. They are not brands. They are bodies.

RAY:

Why does top-down planning still dominate? Master plans. Mega-projects. Zoning reforms.

JACOBS:

Because they comfort the powerful. They make chaos look conquerable. But cities are not problems to be solved. They are mysteries to be stewarded. Plans should be porous. Flexible. Accountable.

RAY:

What's the alternative?

JACOBS:

Conversation. Observation. Incremental growth. And above all: trust in the people who actually live there.

RAY:

What would your message be to today's architects, planners, and politicians?

JACOBS:

Stop drawing and start walking. Talk to people who sit on stoops. Ask what they love, what they fear, what they need. Preserve the mess.

Protect the margins. Honor the old. Embrace the weird. Fight for diversity of use, age, and income. Don't make cities efficient. Make them alive.

RAY:

And to the rest of us?

JACOBS:

Notice. Participate. Push back. Because if you don't shape your city... someone else will.

RAY:

Jane Jacobs didn't just change how cities look. She changed how we see them. She showed us that vibrant neighborhoods can't be engineered from blueprints or summoned from data sets. They grow from trust. From memory. From conflict, connection, and care. In a world obsessed with optimization, she whispered something radical: "A city is not a machine. It is a dance."

Reflection and Commentary: Pedestrians, Not Planners

Jane Jacobs reminds us that cities are not primarily designed by planners – but by pedestrians. She believed in what was seen at ground level: the rhythm of the sidewalk, the give-and-take of strangers, the chance encounter that becomes a friendship.

In the algorithmic city of today, her values sound almost radical. Community over computation. Observation over optimization. In a time when tech promises frictionless futures, Jacobs argued for friction: the beautiful, sometimes aggravating, always human mess of real life.

Her legacy isn't just one of preservation – it's participation. We don't inherit livable cities. We make them, every day, with our attention, our resistance, and our presence.

Further Reading and Exploration

Primary Works
Jacobs, J. (1961). *The Death and Life of Great American Cities*. Random House.
Jacobs, J. (1969). *The Economy of Cities*. Random House.
Jacobs, J. (1984). *Cities and the Wealth of Nations*. Random House.

Modern Commentary and Biographical Sources
Kanigel, R. (2016). *Eyes on the Street: The Life of Jane Jacobs*. Knopf.
Allen, M. (1997). *Jane Jacobs and the Future of New York*. Princeton Architectural Press.

Adjacent Readings
Mumford, L. (1961). *The City in History*. Harcourt.
Gehl, J. (2010). *Cities for People*. Island Press.

25: Aldous Huxley – The Age of Engineered Pleasure

"You are not free. You are distracted."

Preface: Entertaining Control

When we think of dystopia, we often imagine Orwell's jackboot: surveillance, censorship, control through terror. But Aldous Huxley offered a different nightmare – one where oppression wears a grin.

In *Brave New World* (1932), Huxley imagined a world pacified not by fear but by pleasure. A world where citizens are kept docile through engineered happiness, endless distractions, and a little pill called soma. His warning wasn't about totalitarianism. It was about triviality. A society so inundated with comfort, entertainment, and consumerism that it forgets what it means to be human.

Fast-forward to 2025. We live in the dopamine economy. Algorithms serve what we crave. Discomfort is avoided, curated, anesthetized. Huxley saw it coming – and we brought him forward to see what we've done with his prophecy.

Fictional Interview: Aldous Huxley in 2025

RAY:

When people talk about dystopias, they usually mention Orwell. Big Brother. Censorship. Brutality. Control through fear. But what if the future didn't come with a jackboot? What if it came with a smile?

That was the warning of Aldous Huxley. Born into a famous English intellectual family in 1894, Huxley was equal parts mystic and satirist. He saw a world hurtling toward technological advancement without asking what it meant to be human. And in 1932, he published his vision of the future: *Brave New World*.

In it, people aren't terrorized. They're pacified. No gulags. No secret police. Just a population kept docile through pleasure. Through sex without love. Work without meaning. Entertainment without end. And a little pill called soma, to smooth out any inconvenient emotion. It wasn't a nightmare of pain. It was a dream of comfort – engineered to keep people from waking up.

While Orwell feared a world of surveillance and force, Huxley feared a world where we'd stop caring, because we'd be too entertained to notice what we'd lost. He saw consumerism, mass media, and pharmacology converging – not to oppress us, but to soothe us into submission.

Now fast-forward to 2025. We have dopamine loops. Mood playlists. Smart drugs. Algorithmic feeds designed to know what you crave before you do. An entire economy built not just to sell products – but to keep you pacified, optimized, and scrolling.

We don't need telescreens. We don't need whips. We have soma by algorithm.

So we brought Huxley forward. We showed him your screen time. Your open tabs. Your push notifications. And your 14-second attention span. He raised an eyebrow and said: "You are not free. You are distracted."

Aldous Huxley, welcome to the 21st century. Any thoughts?

HUXLEY (light, amused British cadence):

You've built a civilization I imagined – as satire. You've turned my warning into a product roadmap.

RAY:

We call it lifestyle optimization.

HUXLEY:

I called it the final revolution – a world in which people love their servitude. And now they do. Willingly. Ecstatically.

RAY:

In *Brave New World*, people took soma – a drug that removed pain, grief, and thought. Today, we have phones. Social media. Personalized escapism. Same effect?

HUXLEY:

More efficient. Soma dulled awareness. Your tools replace it. You don't feel pain, because you're never alone with your mind. You swipe. Scroll. Stream. Forget. The device is not in your hand. It's in your identity.

RAY:

Orwell feared censorship. You feared distraction. In 2025, what's more dangerous?

HUXLEY:

Distraction is the deeper tyranny. When people are denied information, they may resist. But when they're buried in trivia, they stop caring. You are not forbidden from knowing. You are simply too entertained to notice.

RAY:

We call that "being chronically online."

HUXLEY:

And yet, rarely conscious.

RAY:

We now have algorithms that predict what we want – before we want it. They recommend, suggest, stimulate. Are they our friends?

HUXLEY:

They are our domestication. You are no longer choosing. You are being trained. You live in a pleasure loop – each input optimized for reaction, not reflection. This is not autonomy. It is managed docility.

RAY:

Even Spotify?

HUXLEY:

Especially Spotify. You can't rebel if your soundtrack is soothing.

RAY:

In *Brave New World*, promiscuity was normalized, intimacy was pathologized, and babies were manufactured. Sound familiar?

HUXLEY:

You've created platforms where sex is gamified, connection is transactional, and even heartbreak is content. You don't fall in love. You optimize compatibility.

RAY:

We call that "dating apps."

HUXLEY:

You have replaced vulnerability with velocity.

RAY:

You imagined a world where people felt "happy" all the time – but had no meaning. What's the danger of synthetic happiness?

HUXLEY:

True happiness requires struggle. Complexity. Consciousness. The sanitized version is comfort without consequence. When society promises happiness without discomfort, it must suppress the full range of being human.

RAY:

So, what are we left with?

HUXLEY:

Numb smiles. Curated feeds. And an existential void beneath the emoji.

RAY:

Let's talk about AI. Some say it'll save us from ourselves – solve our problems, optimize society. Sounds utopian?

HUXLEY:

Sounds like the perfect tyrant: benevolent, efficient, and immune to rebellion. When machines define your needs and design your satisfaction, there is no space left for soul. You will be happy – but hollow.

RAY:

In *Brave New World*, there's no war, no poverty, no famine. It's peaceful – but shallow. Would you trade liberty for comfort?

HUXLEY:

No. Because without freedom, comfort becomes confinement. And without truth, peace becomes paralysis. The absence of suffering is not the presence of meaning. Your world has mistaken anesthesia for utopia.

RAY:

How do we escape this – if we even want to?

HUXLEY:

You begin by reclaiming discomfort. Boredom. Silence. Dissonance. Turn off the feed. Go for a walk – without headphones. Ask yourself: what am I avoiding? If pleasure is your only compass, you'll end up very far from yourself.

RAY:

Before we close, I want to bring up someone who admired your work – Neil Postman. In his book *Amusing Ourselves to Death*, he said we feared Orwell's world, but we were living in yours. Thoughts?

HUXLEY:

I'm flattered. Postman understood that the danger was not censorship, but *clutter*. Not tyrants who burned books – but societies that stopped reading because it wasn't fun. He saw the real enemy as triviality – a culture where discourse collapses into entertainment. You don't need a Ministry of Truth when the truth is irrelevant.

RAY:

So, if Orwell warned about a boot stamping on a human face forever…

HUXLEY:

…I warned about a population too amused to notice. Postman extended that insight. He saw your television age clearly. I suspect he'd be horrified by your algorithm age.

RAY:

We interviewed him in Episode 1.

HUXLEY:

Then I hope you listened. Postman is what happens when a culture still has the courage to ask, "What is this doing to our minds?"

RAY:

That was Aldous Huxley. He told us that the future wouldn't need barbed wire. It would offer pleasure – and we'd take the bait.

Reflection and Commentary: Internal Surrender for Frictionless Life

Huxley's dystopia didn't predict violence. It predicted passivity. His world of synthetic pleasure, social engineering, and trivial distraction now seems chillingly familiar.

We often look to Orwell to describe the modern surveillance state. But it is Huxley who saw the deeper danger: that we would willingly surrender our attention, our depth, and even our meaning – for comfort. For pleasure. For another hit of dopamine.

If Orwell warned of external control, Huxley warned of internal surrender. We must ask: what are we trading away in our pursuit of frictionless life? Are we building tools to enrich ourselves – or to numb ourselves?

Further Reading and Exploration

Primary Works
Huxley, A. (1932). *Brave New World*. Chatto & Windus.
Huxley, A. (1954). *The Doors of Perception*. Harper & Brothers.
Huxley, A. (1958). *Brave New World Revisited*. Harper & Brothers.

Modern Commentary and Biographical Sources
Bedford, S. (1973). *Aldous Huxley: A Biography*. Holt, Rinehart and Winston.
Meckier, J. (2006). *Aldous Huxley: Satire and Structure*. Chatto & Windus.

Adjacent Readings
Orwell, G. (1949). *Nineteen Eighty-Four*. Secker & Warburg.
Dick, P. K. (1968). *Do Androids Dream of Electric Sheep?* Doubleday.

26: What Now? A Manifesto for the Living

We've heard from the ghosts. We've summoned the poets, the prophets, the critics, and the code-breakers.

Postman with his TV lament. Shelley with her synthetic monster. Blake, dreaming of angels in a world run by dashboards. Philip K. Dick, wondering if your toaster's gaslighting you.

And through it all, one truth has emerged: they weren't trying to predict the future – they were trying to stop it. They tried to warn us.

So now it's our turn.

Here's what we can do – not to preserve nostalgia, but to reclaim something human in an increasingly posthuman world. A manifesto for the living, from the dead who tried to wake us.

1. Reclaim the Right to Be Confused

"The illusion of knowing is more dangerous than ignorance." – Richard Feynman

We've been taught to fear confusion. To tap, swipe, and scroll our way to certainty. But real thought begins where the UX ends. Confusion is not a bug. It's the birthplace of resistance. Don't fear it. Sit with it. Grow it.

2. Teach Philosophy Before Python

"He who builds the tools must first know the soul." – Hypothetical quote from every ghost in this book

We don't need fewer engineers. We need more ethically literate ones. Make every coder read Foucault. Every startup founder read Morrison. Every AI ethicist touch grass. Because intelligence without introspection is just acceleration.

3. Build Transparent Technology

"The machine should have windows, not mirrors." – Sydney J. Harris

We have built black boxes that run our lives. The answer isn't to smash them – it's to light them up. Demand technology that explains itself. Tools that show their seams. Algorithms with disclaimers. Interfaces that don't seduce, but reveal.

4. Protect the Analog Sacred

"Not everything sacred comes with a subscription model." – Mary Shelley, probably

We must guard the places – and the moments – where surveillance cannot reach. Libraries. Forests. Friendships. Evenings spent doing nothing. These are not anachronisms. They're resistance bunkers.

5. Stop Worshipping Scale

"More isn't better. It's just more." – Ursula K. Le Guin

Not every solution should scale. Some should stay small. If your local food co-op works, don't IPO it. If your art resonates with ten people, let that be enough. The obsession with scale flattens everything sacred into spreadsheet logic.

6. Make Slowness Sexy Again

"You are not falling behind. You are just not being monetized fast enough." – Jim Morrison, after a nap

Slow food. Slow thinking. Slow friendships. The revolution will not be livestreamed – it will be slow-roasted, handwritten, and too boring for the algorithm. Good. That means it's working.

7. Question the Defaults

"Freedom isn't choosing between two apps. It's imagining a third thing." – Philip K. Dick

Every design is a decision. Every platform has politics. Learn to spot the default – and then change it. Or, at the very least, ask: Who chose this for me? And why?

8. Choose Meaning Over Metrics

"You have infinite data. But zero direction." – William Blake, glaring at a KPI dashboard

The goal is not engagement. Or throughput. Or reach. The goal is meaning. What you build. What you say. What you protect. Metrics will always be easier to measure than morals – but they will never be more important.

9. Regulate Like the Future Depends on It

Because it does. Tech doesn't have to be a runaway train. We can build brakes. Regulate surveillance. Tax the bots. Break up the behemoths. Require audits for algorithms. Fund public interest tech. Democracy requires rules. So does dignity.

10. Talk to Each Other Like Humans

"The soul of democracy is the conversation." – Hannah Arendt, from somewhere off-mic

Turn off the feed. Step outside the bubble. Speak face-to-face. Argue in good faith. Touch the ground. Listen without recording. Nothing in this book matters if we forget how to be together.

Final Words: Become the Ghost

This book is not just a warning. It's an invitation. Every person you've heard from – dead or nearly so – was trying to shake us awake. They weren't prophets. They were teachers. And now the classroom is on fire.

You don't need to be a famous thinker to join the séance; just pay attention. The future is not an inevitability. It is a negotiation. And the most radical thing you can do right now – more than voting, posting, or programming – is this: Protect what makes us human. And if the machine asks for your soul in exchange for convenience? Say no.

Then write it down. Pass it on. And when the next generation wakes up in the year 2100 to figure out what the hell we were thinking, may they find your words and whisper: They tried to warn us.

About the author

Ray Welling, PhD, was born and raised in the American Midwest but has lived most of his life in Australia. He holds a BSJ from Northwestern University's Medill School of Journalism in Chicago, an MA in mass communication from Macquarie University and a PhD in marketing and management from the University of Sydney. Following a professional career as a journalist, publisher, content marketer and creator, strategy consultant, digital marketing manager and writer, Ray turned to academia, and has taught digital marketing, consumer behaviour, public relations, social media, branding, creative advertising, media/marketing convergence, strategic communications, global marketing and decision-making to undergraduate, postgraduate and MBA students at several Australian universities over the past 15 years, as well as conducting research into website success measures, social media and ethics. He is the author of *Digital Disruption and Transformation: Lessons from History* (2018) and *Byline for the Dead* (2025), and hosts the Clear as Mud and They Tried to Warn Us podcasts.

www.ingramcontent.com/pod-product-compliance
Lightning Source LLC
Chambersburg PA
CBHW011150290426
44109CB00025B/2556